BIOLUMINESCENCE

BIOLUMINESCENCE

Living Lights, Lights for Living

Thérèse Wilson

J. Woodland Hastings

Harvard University Press
Cambridge, Massachusetts
London, England
2013

Copyright © 2013 by the President and Fellows of Harvard College

All rights reserved

Printed in Canada

Library of Congress Cataloging-in-Publication Data

Wilson, Therese, 1925–
  Bioluminescence : living lights, lights for living / Therese Wilson,
J. Woodland Hastings.
      p.   cm.
  Includes index.
  ISBN 978-0-674-06716-5 (alk. paper)
  1. Bioluminescence.   I. Hastings, J. Woodland (John Woodland), 1927–.
II. Title.
  QH641.W55 2013
  572'.4358—dc23   2012022227

CONTENTS

ACKNOWLEDGMENTS

It is our pleasure to acknowledge our debt to the many who have been critical to the birth of this book. Foremost among those are the students, postdoctoral fellows, and visitors to the Hastings and Wilson lab, who, over the years, created a unique and unforgettable atmosphere of free exchange of ideas, and joy in exploring all sorts of topics related to the emission of light by living (and non-living) things. Our debt to all of them cannot be expressed adequately.

We are especially grateful to Jim Morin, who read the entire manuscript and pointed out its weak points, and to Anne Goldizen for her editorial suggestions throughout the manuscript. Many friends did us the favor of reading chapters or parts of chapters. Their comments and suggestions were universally appreciated. Among those to whom we are especially grateful are Etelvino Bechara, Arthur Halpern, Andy Knoll, Neil Krieger, Ken Nealson, John Paxton, Deb Robertson, Vadim Viviani, and Edith Widder. Thérèse Wilson is especially grateful to Meredithe Applebury for her advice and encouragement at the very early stages of the work.

Without illustrations, a book like this would not see the light of day. We thank all those who have given us permission to use photos and illustrations, especially Milton Cormier, Larry Fritz, Helen Ghiradella, Karsten Hartel, Peter Herring, and Christopher Kenaley. Thanks also to Marvin Morales for the chemical formulae. For all other figures, we were exceedingly lucky to be able to rely on Renate Hellmiss, whose artistry, competence, and patience made working with her a pleasure at all stages of the work.

Finally, we are grateful for financial help from the Faculty of Arts and Sciences and the Department of Molecular and Cellular Biology of Harvard University.

BIOLUMINESCENCE

Figure I.1. A stretch of the Southern California coastline at the time of a "red tide," a bloom of luminous dinoflagellates, *Lingulodinium polyedra*. The blue streak is the bioluminescence of cells stimulated to emit by wave action on the beach.

INTRODUCTION

For many of us, the only awareness of bioluminescence—the emission of light by living organisms—is the sight of fireflies. For others, the magic moment is a "phosphorescent sea," where myriads of microscopic sources emit flashes of light when waves or our steps on wet sand stir the water (Figure I.1). In truth, bioluminescence is everywhere on earth, especially in the oceans, from the depths where no sunlight penetrates to the surface, where it powers photosynthesis. Emission of light is a function that was invented, reinvented for different reasons, with different chemistries and in different species, and perhaps lost countless times during the course of evolution. There is no way to tell.

On a tree of life (Figure I.2), the scattered red dots make this point. Many mushrooms are bioluminescent, but no plants; many insects emit light, but no spiders. Among vertebrates, bioluminescence is found abundantly in fish, but not in frogs, salamanders, birds, reptiles, or mammals. Today's reptiles, birds, and mammals communicate by sounds, odors, touch, and visual signals. Why did so many species fail to acquire the ability to emit light, or to take advantage of a symbiosis with bioluminescent bacteria, for example, as many fish and squid did?

The emission of visible light (Figure I.3) requires energy. We know how hot a light bulb has to get before emitting light, yet evidently a firefly does not fry. This is because the mechanisms of light production in the firefly and in the light bulb are different. The current passing through the bulb's filament heats it, and this thermal energy is then radiated as light. The hotter the filament, the more intense and white rather than red is its incandescence, "white-hot" and "red-hot" in blacksmiths' talk. Within the lantern of the firefly, a biochemical reaction is the source of the emission. A specific molecule is generated in a high energy state, out of thermal equilibrium with its surrounding; this "excited" molecule immediately radiates

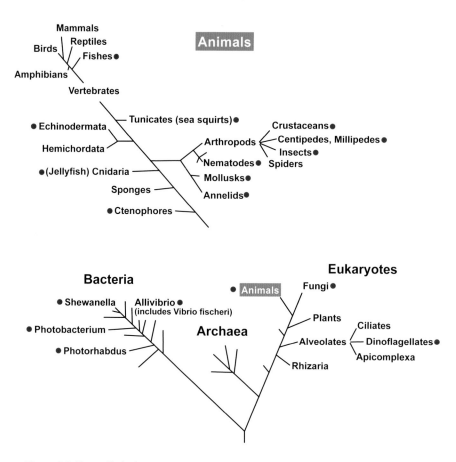

Figure I.2. Tree of Life (in two connected parts). The scattered red dots mark the taxa in which bioluminescence has been observed. Not all species in a red-labeled taxon are luminescent, far from it. Of more than 10,000 known millipede species, for example, only 8 have been found to emit light, while fewer than 100 of the 1.5 million mushroom species have been described as luminescent to date.

its surplus energy as light, usually within a nanosecond (a billionth of a second), and very little heat is produced. Bioluminescence is of great interest because, despite its being part of complex biological systems, the light itself results from a single, very specific step in a biochemical reaction.

All bioluminescences are without exception reactions of oxygen (O_2) with molecules specific to a particular group of organisms, catalyzed by specific enzymes. All involve the production of transient

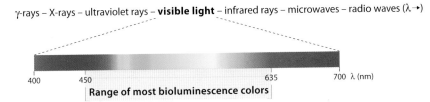

γ-rays – X-rays – ultraviolet rays – **visible light** – infrared rays – microwaves – radio waves (λ→)

400   450   635   700 λ (nm)

Range of most bioluminescence colors

Figure I.3. The dual nature of light. Light may behave either as a stream of particles of energy (photons) or as electromagnetic waves that can be dispersed by a prism to produce a spectrum—a rainbow, as drawn here. The visible part of the electromagnetic spectrum is sandwiched between ultraviolet and infrared (some animals see further in the UV, some further in the IR than humans). When we focus on biological damage caused by light, as in sunburn, or on the emission of light by a biochemical reaction, as in bioluminescence, we think of light as a stream of photons. Each photon carries a specific amount of energy, equal to $h\nu$, where h is a constant and ν is the frequency of the light, thus inversely proportional to its wavelength. Photons of γ-rays carry ca. 10^{20} more energy than radio-wave photons, and photons of violet light carry as much as four times more energy than red photons. Most bioluminescences peak between 450 and 635 nm, as indicated; however, some deep-sea fish emit at wavelengths down to 400 nm, others up to 700 nm. Photons of green light, often the color of bioluminescence, carry packets of energy of the same order of magnitude as that binding two carbon atoms in a molecule.

peroxides, where the O-O bond between the oxygen atoms is weak and easily broken. This step generates a product with, by contrast, a strong carbon-oxygen bond. It is the energy released by this type of reaction that brings a product to an "excited state," from which the photons of bioluminescence are emitted when the product falls back to its lower, "ground state" level. How chemical processes can lead to the emission of light is discussed in Chapter 11, which some may choose to read first.

What makes bioluminescence so fascinating is that diversity is found at all levels, from the organisms themselves and the use they make of the light they emit—defense, offense, communication—to the anatomy of the light organs and the biochemical reactions ultimately emitting the light as photons. It is, therefore, a topic that can be approached from many angles.

In Part I, we look at organisms where the mechanism of light emission is understood at the molecular level. We first describe a small marine crustacean, *Vargula*, in which the light-emitting system is a textbook example of the kind of chemistry that is involved. We then turn to two coelenterates with closely related chemistries of emission, *Renilla* and *Aequorea*, the latter a poster child of bioluminescence. Our next topic takes us to dry land with fireflies and other beetles, how they produce light, and what they use it for—courtship, for sure, but in some cases also for catching a meal. In the next two chapters, the focus is on two very different single-cell marine organisms, dinoflagellates and bacteria. In the latter chapter, we describe various beautiful and strange bioluminescent fishes and a remarkable little squid that is teaching us how symbioses work.

Part II starts with a collection of short stories describing a dozen or so fascinating bioluminescent creatures that we know less about, from the lurid fireworm of Bermuda to a shy little fresh water snail of New Zealand to nasty bacteria that sometimes infect wounds but also collude with tiny worms (nematodes) to kill insects, a textbook example of mutualistic gangsterism. We visit the bioluminescent earthworms and mushrooms, which are perhaps best for considering the question: what on earth are they gaining from emitting light? There are ideas, but few experiments to test them.

The next topic of Part II considers bioluminescent organisms living in the oceans. They far outnumber their terrestrial counterparts, which, to be sure, are restricted to a space on earth only some one-hundredth of that of the oceans' volume. Moreover, and remarkably so, luminous organisms are essentially absent in fresh water, and a fully satisfactory explanation for this has never been found. It surely lies in the realm of evolution, which will be discussed.

In oceans that can be miles deep, sunlight penetrates significantly only about 200 m below the surface and becomes increasingly blue with depth. It is in the top layer of the ocean, where photosynthesis is at work and life is abundant, that many of the marine bioluminescent creatures we will talk about live.

Below is a dark blue twilight zone, and from 1000 m down darkness is constant and total, except for bioluminescence. Apart from the environment of hot vents, where a special fauna thrive but no bioluminescence has been reported, any life below 1000 m owes its existence to the nutrients that slowly drop down to the ocean floor. This immense dark, cold, and sparsely populated world (someone calculated that there is only one female anglerfish per cubic mile in this zone) is where some of the most outlandish bioluminescent fishes cruise for a living.

In the next chapter, we propose that the functions of bioluminescence in different organisms all fall under four major headings: defensive, to escape predators; offensive, to aid in predation; communication; and propagation. In the last chapter of Part II, we discuss the origin of bioluminescence. Its survival value today is certainly based on its functional importance, but billions of years ago, when oxygen first appeared on earth as the result of photosynthesis, it would have been toxic to all organisms, which had been living in an atmosphere lacking it, and the enzymes of bioluminescence may have played a critical role in oxygen removal.

The study of bioluminescent species has opened a treasure trove of exquisite tools for biological research, examples of which are highlighted in Part III. Fireflies, luminous bacteria, crustaceans, and jellyfish, among others, have given us enzymes, substrates and fluorescent proteins that are now used all over the world in all fields where molecular biology and imaging techniques are applied. On the practical side, we go to firefly or bacterial bioluminescence, for example, to check that our meat or our drinking water is not contaminated.

At the same time, but at a deeper and more fundamental level, the study of luminous bacteria has taught us mechanisms by which individual bacteria chemically communicate with and count others of their kind, and regulate the expression of specific genes accordingly. This is a discovery of fundamental importance for how we look at bacteria more generally. The basic mechanism of such cell-to-cell communication, dubbed "quorum sensing," is now understood and shown to play a critical role in diseases such as cystic fibrosis, for example, as well as in microbial biofilm formation and some fascinating symbioses.

Bioluminescence was once thought of as merely an esoteric area of research. How it has opened doors to many fields of discovery is a story that we hope to bring to life here.

part one

FIVE DIFFERENT BIOLUMINESCENCE SYSTEMS

Sixty years ago, when E. Newton Harvey of Princeton University wrote his masterly book *Bioluminescence*, which remains today, with its sixteen-page index, the foremost source of information on the countless animals that emit light, little was truly known on how they accomplish such a feat. The critical role of oxygen was perceived, although not as uniquely as we now know it to be, and so was that of enzymes (the special catalytic proteins called luciferases) and the smaller molecules (luciferins) whose reactions with oxygen the luciferases catalyze.

Today, it can be said that in several cases the exact step has been established where the reaction of oxygen with an enzyme-bound luciferin produces light emission. Such success stories are where our book starts. Taken together, these first five chapters illustrate the curiously haphazard impact of evolution on bioluminescence, many different animals discovering their own molecules for the emission of light.

As in most fields of biological research today, biochemists, molecular biologists, chemists, and the physicists who invented powerful new techniques and instrumentation were successful by working together. Bioluminescence is a field of study where progress depends as much on the naturalists exploring remote areas at night as on state-of-the-art instrumentation that ultimately reveals the mechanism of the light emission.

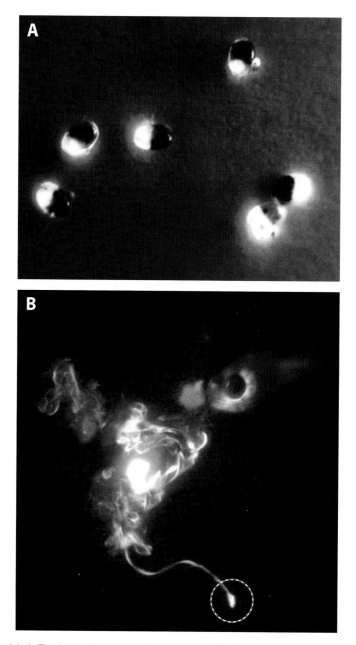

Figure 1.1. *A.* The bioluminescence of several small bioluminescent crustaceans (*Vargula hilgendorfii*) at night. Females of all Vargula species are about 50 percent larger than males. *B.* Massive secretory defensive display response of the luminescent Caribbean cypridinid ostracod *Photeros annecohenae*, swimming away unharmed at the bottom (*circled*), to an attack by the cardinal fish (*Phaeoptyx pigmentaria*) at the upper right. Note the luminescence still glowing within the oral cavity of the fish where the ostracod first responded before being spit out by the fish. The ostracod is rejected even in daylight, which suggests that it is distasteful in addition to being bioluminescent.

chapter one

A MARINE CRUSTACEAN

Bioluminescent Fishes as Plagiarists and Thieves

Vargula hilgendorfii (or *Cypridina hilgendorfii*, as it used to be called) is a small crustacean in the family Cypridinidae* that deserves pride of place in this book for the textbook simplicity of its biolumines-cence mechanism. It is about 2 to 3 mm in diameter, protected by a hard shell (Figure 1.1A), and abundant in shallow waters along the southern coast of Japan, as well as in many other areas.

Buried in the sand during the day, it becomes an active scavenger at night, escaping potential predators by squirting a bright blue lumi-nescent puff, which may startle or temporarily blind them, or may serve as a decoy, allowing their escape in darkness. Sometimes the luminescence may attract a predator of the first predator, which eats it or scares it to regurgitate the cypridinid it had captured. There is also strong evidence that cypridinids are distasteful; after being taken in by a predator, such as a fish, they are regurgitated and may swim away unharmed (Figure 1.1B).

In the Caribbean Sea and, remarkably, apparently not known else-where, luminescent males in this family are more inventive. They produce astonishing pulsed displays that leave punctuated tracks of blue light in the water as courtship signals to females (Figure 1.2). Male *Photeros annecohenae*, for example, start by emitting either two or four pulses at the bottom, at the level of the sea grass. They then swim upward for half a meter in a helical pattern, emitting a light pulse every other turn. If this display does not attract a female, the male goes straight back down and starts spiraling up again. The timing precision of these displays matches that of firefly flashes. Around some reefs in the Caribbean Sea, the local male cypridinids

---

* About half the 400–500 species of this family are luminescent.

Figure 1.2. Courtship displays created by beads of bioluminescence squirted into the water by male cypridinids, swimming at a speed greater than 3 cm per second (or about twenty body lengths a second!); the displays are visible from at least 15 meters. In attempts to attract sexually receptive females, horizontally displaying males competing concurrently leave trails of pulses in complex fan-like patterns (*upper left*) above the reef. A starlight camera recorded the luminescent displays simultaneously with an infra-red camera recording the reef (made visible by means of infrared lights). Both horizontal and vertical display trains are visible, produced by two new but still unnamed species.

Figure 1.3. Horizontally displaying male cypridinid ostracods (a new species in a new genus, name not yet assigned) that produce displays above reefs in Belize, such as shown in Figure 1.2. Each male is 1.6 × 0.95 mm and has a pair of large black compound eyes. The light organ is the yellowish bar below the eye that extends toward the notch at the anterior end of the valves.

(Figure 1.3) favor horizontal displays, but again with the same timing precision.

Cypridinids secrete their bioluminescence reagents from two separate glands, as a chemist would do from two test tubes. Both secretions are remarkably stable as dry powders. During World War II, the Japanese collected and dried vast amounts of *Vargula hilgendorfii*; they thought that by moistening the powder on the palm of their hands or spreading it on each other's backs, soldiers would be able to read maps or keep track of one another at night. Such clever tricks were apparently never employed, but the large store of dry *Vargula* came to good use later for elucidating the chemistry of the reaction.

How does it work? This is the question that observers of bioluminescences, now as in centuries past, ask first, fascinated as they are by the mystery of a firefly flash or the puff of blue light in the ocean. *Like all bioluminescences, emission from* Vargula hilgendorfii *is the result of a reaction of oxygen (O$_2$),* the form of oxygen that we breathe. No exception to the rule that oxygen or another of its active forms is required for bioluminescence has ever been reported. The oxygen requirement had already been demonstrated in the seventeenth century by Robert Boyle, who observed that the luminescence of decomposing fish (now known to be caused by bacteria) or of fungi on rotting wood was extinguished if air was pumped out but restored when air was readmitted. However, the specific molecules with which oxygen reacts differ from one bioluminescent group of organisms to another; this is a key feature of bioluminescence, important to keep in mind.

At the end of the nineteenth century, the French scientist Raphael Dubois established that both a heat-sensitive component (an enzyme that he called *luciferase;* see Glossary) and a much smaller component (the substrate called *luciferin*) play a role in firefly (beetle) bioluminescence. In fact, all bioluminescence reactions are catalyzed by enzymes that are generally specific to each taxon. In shorthand, such reactions can be expressed as follows:

$$\text{Enz-luciferin} + O_2 \rightarrow \text{Enz-oxidized luciferin*} + CO_2$$
$$\text{Enz-oxidized luciferin*} \rightarrow \text{Enz} + \text{oxidized luciferin} + h\nu$$

These two lines tell that luciferin binds with luciferase (Enz) and then reacts with oxygen; one might say that the role of binding is to put the luciferin in a proper mood to react with oxygen.

The *asterisk* on "oxidized luciferin*" indicates that this reaction product is in a so-called excited state, a state richer in energy, from

Box 1.1. What is fluorescence?

We know that some materials "fluoresce" when exposed to light; that is, they take in the energy of "exciting" photons (hv_{EXC} in figure below) that are usually present in ambient light and re-emit some of their energy as fluoresence light (hv_{FL}), whose color is shifted to the red of that of the exciting light. If one exposes tonic water to ultraviolet light, the quinine that it contains emits a beautiful blue fluorescence; in shorthand this is written in two steps:

$$Q + hv_{EXC} \rightarrow Q^*$$
$$Q^* \rightarrow Q + hv_{FL}$$

Fluorescence is shifted to the red of the exciting light, because some of the energy of the exciting photon ends up inducing motions of vibration and rotation in the newly excited molecule, and therefore hv_{FL} contains less energy than hv_{EXC}. Thus, fluorescence spectra are moved to the red of the absorption spectra. One can think of two levels, the normal, ground-state level, and the excited-state level, with vertical arrows up and down representing absorption and emission of photons, as shown in the figure below.

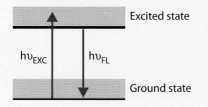

Excited state

Ground state

Diagram of light absorption and emission in fluorescence. The gray clouds in the ground and in the excited states represent vibrational and rotational levels associated with the two electronic states.

which it promptly falls back to the "ground state" with the release of its excess energy, either as light (a photon, designated by hv) and/or as heat, if the conversion is not 100 percent efficient—it never is. The way to grasp how this works is to remind ourselves of fluorescence (Box 1.1).

The cypridinid reaction is *the simplest known* bioluminescence system, a textbook example of bioluminescence chemistry. One of the two secretions released in the seawater contains the luciferin, called cypridinid luciferin (so as to include all members of the Cypridinidae

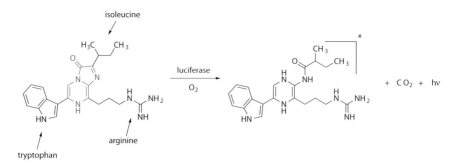

Figure 1.4. The chemical structure of the cypridinid (*Vargula*) luciferin and of the product of its reaction with oxygen. Note the red central structure of luciferin (the imidazopyrazinone), which is also found in coelenterates' luciferin, albeit with different amino acids substituents (see Chapter 2).

family), whose structure has been established (Figure 1.4); the other is the luciferase. *It is the reaction of oxygen with luciferin that provides the energy emitted as light.*

When the two secretions come in contact in the seawater, luciferase binds luciferin in a conformation favorable to its reaction with oxygen. Dissolved oxygen then reacts with it at a very specific location in the molecule. The reaction generates carbon dioxide (CO_2) and oxyluciferin in its excited state, from which the blue bioluminescence is emitted.

Where does the energy carried off by the blue photons come from? Remarkably, this question has been answered unequivocally. Oxygen reacts with luciferase-bound luciferin to form an unstable peroxide, a molecule with two oxygen atoms attached to each other at the corner of a square structure, called a dioxetanone. It is the breakdown of this unstable peroxide that generates the energy for emission (Box 1.2; see also Chapter 11).

Luminous Fishes with Chemistries Kindred to Crustaceans

Along the U.S. Pacific Coast, *Porichthys*, also called the midshipman fish (Figure 1.5A), emits light during the mating season from the hundreds of aligned photophores located ventrally, which give

Box 1.2. The bioluminescence reaction of *Vargula hilgendorfii* and other cypridinids.

As described, the reaction of luciferase-bound cypridinid luciferin with oxygen forms a peroxide, which rapidly cyclizes to form a transient cyclic peroxide, too unstable to isolate; its breakdown is the source of the energy required to generate the product oxyluciferin in its excited state, hence the bioluminescence.

The intermediacy of the transient dioxetanone shown below (between parentheses) was established by an elegant trick available in chemists' toolbox, that of using isotopically labeled "heavy" oxygen, $^{18}O_2$, instead of "normal" oxygen, $^{16}O_2$. If the reaction proceeds as shown above, every molecule of CO_2 generated ought to carry one atom of ^{18}O, the "heavy" oxygen, and one atom of ^{16}O, the "normal" oxygen. This is indeed what the results of the experiment established by mass spectral analysis. The bioluminescence spectrum matches the fluorescence of oxyluciferin in the presence of luciferase, in accord with this mechanism. The efficiency of the reaction, its "quantum yield," is high: 30 percent of the enzymatically oxidized luciferin molecules emit a blue photon.

Cypridinid luciferase is a ~62 kDa protein. It has been cloned and sequenced, revealing two glycosylation sites. Two sugar chains are necessary for activity; they have been taken advantage of in an application involving attaching a dye to the luciferase, which is itself targeted to cancer cells (see Part III, Bookends). But the sugar chains have so far prevented the crystallization of the luciferase and the establishment of its three-dimensional structure. The N-terminal sequence is consistent

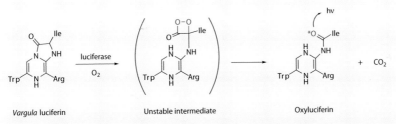

Vargula luciferin   Unstable intermediate   Oxyluciferin

The substrate, intermediate cyclic peroxide, and excited product in the cypridinid reaction.

with a secreted protein. Because the expressed protein can be assayed nondestructively in the medium outside the cell, the gene of cypridinid luciferase has proven useful as a reporter for many assays.

Cypridinid luciferin was obtained as pure crystals in 1957; in 1966, its chemical structure as an *imidazopyrazinone* was elucidated, and it was synthesized in the laboratory. The structure suggests that cypridinid luciferin is biosynthesized from three amino acids, tryptophan, isoleucine, and arginine. Feeding experiments support this hypothesis, although the enzymes involved in the biosynthesis have not yet been isolated.

it its name (Figure 1.5B). The male buries himself in the sand during the day, and swims just above it at night. As if its light display was not enough to seduce females, it also sings during the mating season, in a true display of *Son et Lumière*. The loud noise is known to sometimes disturb residents of houseboats in the San Francisco Bay.

Each of the photophores contains both cypridinid-type luciferin and a luciferase, which cross-react to give light with authentic cypridinid luciferase and luciferin, respectively. Does the fish synthesize its own luciferin? Probably not. A fascinating finding is that midshipmen fish from the Puget Sound area fail to emit light, even though they have fully developed photophores and a luciferase. The problem was traced to their diet. It was found that these fish become luminous after ingestion of a luminescent cypridinid species or are injected with purified cypridinid luciferin. This can be compared to a vitamin in humans, although it is needed for light emission in this case rather than life itself.

While it is evident that the Puget Sound fish cannot synthesize luciferin under their normal nutritional conditions, they do synthesize an active luciferase, although it is not known if this luciferase is chemically the same as authentic cypridinid luciferase.

Parapriacanthus beryciformes is another example of a fish that has co-opted the cypridinid system in its two big light organs (Figure 1.6).

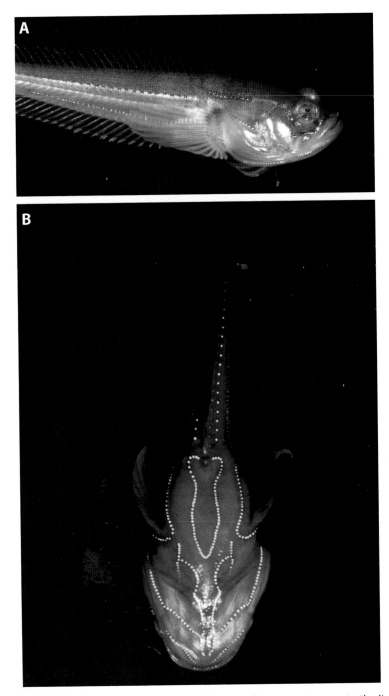

Figure 1.5. *A. Porichthys*, the so-called midshipman fish, owes its name to the linear arrays of photophores along the ventral part of its body. *B.* The ventral photophores visualized by their bioluminescence.

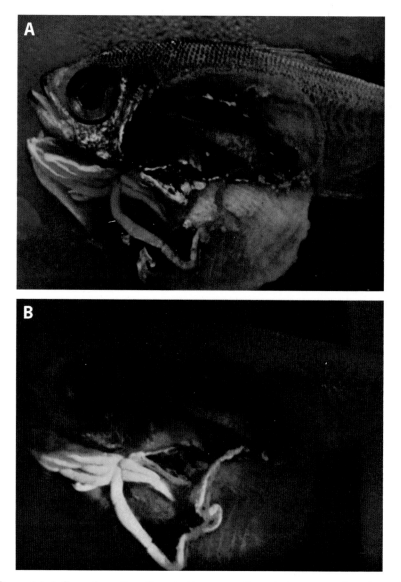

Figure 1.6. *A. Parapriacanthus* dissected to expose the pyloric caeca and portions of the visceral organs containing *Vargula* (cypridinid) luciferin, photographed in room light. *B*. The same fish photographed under ultraviolet light, showing the fluorescence of the luciferin.

In fact, large amounts of cypridinid luciferin and dead cypridinid bodies were found in its digestive system, which supports the conclusion that this fish also does not synthesize it, but ingests it. *Parapriacanthus* may also make its own luciferase, like *Porichthys*, since it develops the organs to accommodate the light-emitting system.

Figure 2.1. The soft coral *Renilla reniformis*. Live specimen, approximately 2 cm in diameter, showing expanded polyps, which are bioluminescent.

chapter two

JELLYFISH AND GREEN FLUORESCENT PROTEIN

A Soft Coral, a Calcium-Sensitive Protein, and

Fish with Related Bioluminescence Systems

From the cypridinid crustacean *Vargula*, we turn to the soft coral *Renilla reniformis*, a coelenterate often called the sea-pansy, which is found in the North Atlantic coastal waters from North Carolina to Florida (Figure 2.1). *Renilla* emits its luminescence from an array of cellular point sources (called photocytes) within its tissues. When touched, a wave of light bursts through the colony, scaring predators.

Remarkably, the luciferin of *Renilla*, called coelenterazine, shares with cypridinid luciferin the critical *imidazopyrazinone* structure (in red), albeit with different substituents (Figure 2.2).

It can therefore be safely bet that the light-generating reactions of cypridinid and *Renilla* luciferins with oxygen follow a similar path. However, the bioluminescent reaction of *Renilla*, as it unfolds in the living organism, is not as textbook-simple as that of cypridinids. It follows a multi-step mechanism that involves other proteins–notably, a luciferin-binding protein that holds and protects the luciferin and releases it only in the presence of calcium, allowing it to then bind with luciferase and react with oxygen, yielding the excited product, called coelenteramide.

An intracellular system allows *Renilla* to control its bioluminescence emission at the biochemical level, and very precisely so, in contrast to cypridinids, which only time the release of luciferase and luciferin in the seawater. In extracts (*in vitro*), the light emitted is blue, with a broad spectrum peaking at 480 nm. But in the animal (*in vivo*), remarkably, the emission is green ($\lambda_{max} = 509$ nm) and its spectrum is narrow (Figure 2.3).

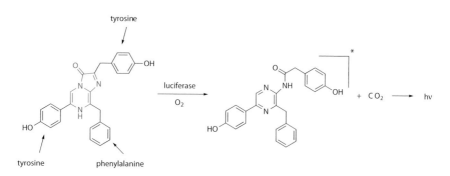

Figure 2.2. *Renilla* luciferin (called coelenterazine) and the excited product of its luciferase-catalyzed reaction with oxygen, called coelenteramide. Note the imidazopyrazinone central structure (*in red*), present also in cypridinid luciferin.

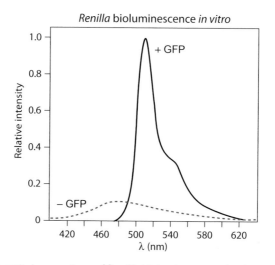

Figure 2.3. The emission spectrum of *Renilla* bioluminescence *in vitro*, with (*solid line*) and without (*dashed line*) added rGFP.

This shift from blue to green is due to the association of *Re-nilla* luciferase with another protein, *Green Fluorescent Protein*, or *rGFP* (r for Renilla). The protein rGFP acquires its energy directly from the excited reaction product, coelenteramide,* by what is called an "energy transfer" step, summarized below. How energy transfer works, important in bioluminescence, is discussed in Chapter 11.

$$\text{Coelenterazine} + O_2 \quad \rightarrow \text{coelenteramide}^*$$
$$\text{coelenteramide}^* + rGFP \rightarrow \text{coelenteramide} + rGFP^*$$
$$rGFP^* \qquad\qquad\qquad \rightarrow GFP + h\nu$$

Renilla is not the only coelenterate to produce a fluorescent protein associated with its bioluminescence. The beautiful jellyfish *Aequorea victoria* (Figure 2.4), which emits green light from distinct structures (photophores) at the places around its umbrella where the tentacles attach, uses the same luciferin as *Renilla*—coelenterazine—and coelenteramide is also the primary excited-state product. However, *Aequorea*, which acquired fame for its own GFP (aGFP), found a fascinating way to poise the system for its quick and ready flash. It makes and stores *a stable luciferase-luciferin reaction intermediate*, called *aequorin*, in which a molecule of oxygen is already covalently bound to the luciferin.

We can look at it this way: the protein part of aequorin holds together and shields from the outside what can be regarded as a stable reaction intermediate, made of the luciferin already bound to oxygen, in turn bound to the protein. If and when calcium ions (Ca^{++}) are added to this complex, they bind to the protein and induce an immediate change of its three-dimensional conformation. This, in

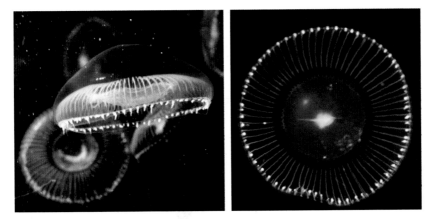

Figure 2.4. The bioluminescent jellyfish *Aequorea victoria* (left), showing the light-emitting photocytes located at the periphery of the umbrella (right).

turn, allows the two peroxidic oxygen atoms, already attached to the molecule of luciferin, to form a short-lived cyclic peroxide, as in the case of cypridinid *luciferin* discussed above. Within milliseconds, this peroxide decomposes and generates excited coelenteramide, hence blue light in the absence of aGFP.

Consistent with this scenario, the crystal structure of aequorin shows networks of hydrogen bonds stabilizing coelenterazine with oxygen bound to it inside the protein. Calcium triggers shifts or breaks in some of these bonds, starting a cascade of rearrangements: the first step is the cyclization of the peroxide group to form a short-lived cyclic peroxide, as in the systems discussed previously.

Starting with aequorin, the stable intermediate, the overall reaction thus goes as follows:

$$\text{Aequorin} + \text{Ca}^{++} \rightarrow \text{apo-aequorin} + \text{CO}_2 + \text{coelenteramide}^* \rightarrow \text{h}\nu$$

What happens to the protein part of aequorin after this reaction is over? Behaving now as a luciferase, if and when coelenterazine and oxygen are present but calcium ions are not, it can very slowly reform aequorin. This slowness ensures that there will be little or no "afterglow," that is, emission resulting from enzyme turnover after the flash. *In vitro*, in the presence of only traces of Ca^{++}, a mixture of coelenterazine and apo-aequorin emits a very low, steady glow. But aequorin can accumulate only in the complete absence of calcium ions. In the photocytes of *Aequorea*, where there is no calcium, aequorin can be regenerated, provided coelenterazine is present.

In one reaction cycle, a so-called single turnover, the amount of light emitted is proportional to the amount of aequorin. Because of its high sensitivity and specificity, aequorin is an *in vivo* probe of choice for intracellular calcium ions, down to the 10^{-6} or 10^{-7} M level. Taken together with its remarkable stability, it is not surprising that aequorin was nicknamed a "photoprotein." Unfortunately, this term brings to mind a light-*activated* rather than a light-*generating* system.

Green Fluorescent Proteins, the GFPs

In vitro and in the absence of GFPs, the reactions of both *Renilla* and *Aequorea* produce blue light, peaking at 485 nm in the case of *Renilla*, and at 460 nm from *Aequorea*, the color difference probably due to different protein environments around the excited coelenteramide molecules.

In the living organisms, however, both *Renilla* and *Aequorea* emit green light, peaking at nearly the same wavelength, ~509 nm (Figure 2.3, *Renilla* rGFP). The gap between the peaks of *in vivo* and *in vitro* emission was attributed, already 50 years ago in the case of *Aequorea*, to a "fluorescent green protein," which could be separated from the luciferase on a chromatographic column. The chemical structure of its chromophore (the specific part of the GFP molecule responsible for the green fluorescence) was identified two decades later as an imidazolone, similar to a part of the chromophores of both cypridinid luciferin and coelenterazine.

Note that, as mentioned, this "fluorescent green protein" came through the chromatographic column *with its chromophore—more precisely, its fluorophore—still attached*. Indeed, the chromophore is an integral part of the protein to which it is covalently bound. When the cDNA of this GFP was originally obtained and cloned in *E. coli*, the expressed protein already contained the chromophore. Remarkably, the complete primary amino acid sequences of *Renilla* rGFP and *Aequorea* aGFP are quite different, which is reflected in their absorption spectra. But their fluorophores are identical, and therefore their fluorescence spectra are almost superimposable.

These two GFPs are extremely stable thermally, for reasons that became clear only when their crystal structures were obtained. Indeed, they show the chromophores sitting in the middle of protective cylindrical barrels formed by the rest of the amino acid chains of their respective proteins (Figure 2.5). Nevertheless, in the cases of both the *Renilla* and *Aequorea*, the excited coelenteramide can come close enough to the GFP chromophores for energy transfer to take place, resulting in green emission.

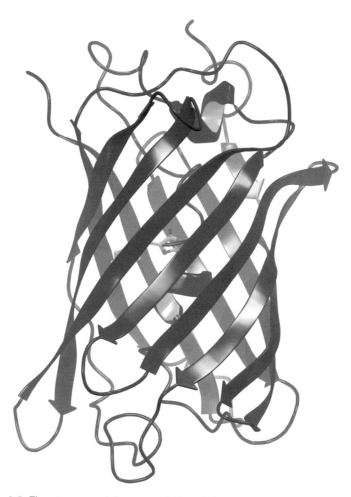

Figure 2.5. The structure of *Aequorea* GFP (aGFP), showing the covalently linked chromophore between the two red markers at the center of the can-like structure. In its protected position, the chromophore is accessible to water only from the top or from the bottom of the can but is out of the reach of an enzyme or large peptide.

This overview of coelenterate bioluminescence has singled out, perhaps unfairly, only two examples, *Renilla* and *Aequorea*. A more extensive discussion would cover other, equally interesting examples, but would not bring up significantly different features. *Renilla* is a fair representative of most anthozoans, while *Aequorea* can be regarded as representing hydrozoans.

We therefore owe to the coelenterates' bioluminescence systems two invaluable gifts: first, their unparalleled sensitivity to calcium ions, which is the basis of the most sensitive method of detection and measurement of intracellular Ca^{2+}, and second, the GFPs, which have become truly ubiquitous in practically every field of biological research, from developmental biology to neurobiology. A Nobel Prize was awarded in 2008 for this work.

Luminous Fishes with Coelenterazine-Based Chemistries

While some fishes, as we saw, use a cypridinid type of bioluminescence chemistry, others co-opt coelenterazine-based bioluminesecence. The dichotomy seems to be that coastal fishes use cypridinid luciferin, whereas coelenterazine is definitely found, sometimes in large amounts, in deep-sea fishes. The liver of one species, *Neoscopelus macrolepidotus* (Figure 2.6), is full of coelenterazine. Whether it acquires it in its diet or synthesizes it is an open question. The same goes for the luciferase.

In any case, whether a fish adopts a bioluminescent system based on cypridinid luciferin or on coelenterazine, it is not an association

Figure 2.6. The large (length up to 22 cm) lantern fish *Neoscopelus macrolepidotus*, with rows of photophores on the belly and along each side.

or symbiosis between the fish and a bioluminescent partner; the fish needs molecules of coelenterazine for bioluminescence, the way mammals need vitamin A for vision. Chapter 5, in contrast, shows examples of fishes whose light organs harbor and maintain populations of symbiotic luminescent bacteria.

The Ubiquitous Coelenterazine

The reader may think that coelenterazine is the luciferin of only *Renilla*, *Aequorea*, and some odd fishes. In truth, it has been identified as the luciferin of many other species, including ctenophores, crustaceans (copepods, decapods, some shrimps), as well as some squids, and it has also been found in quite a number of nonbioluminescent species.

Again, do all these organisms synthesize their coelenterazine? Apparently they do not. Coelenterazine, as we saw, shares its imidazopyrazinone skeleton with the cypridinid luciferin of *Vargula*, differing from it only in its three amino acids substituents. But while *Vargula* synthesizes its own luciferin, coelenterates apparently cannot and must acquire coelenterazine from their diet. An aquarium population of *Aequorea victoria* fed on a coelenterazine-free diet does not luminesce.

The question is, then, what are the organisms, luminous or not, that are able to synthesize coelenterazine and introduce it in the food chain? The first to be reported is *Systellaspis debilis*, or, more exactly, the eggs of this small shrimp. These eggs, once attached externally to the mother shrimp, receive no nutrients from her. While newly laid eggs contain only trace amounts of coelenterazine, by hatching time their coelenterazine content is nearly two orders of magnitude greater, establishing clearly that the shrimp have the capability of synthesizing the compound. With the methodology previously used to demonstrate that *Vargula* synthesizes its luciferin from arginine, isoleucine, and tryptophan, a recent study with deuterium-labeled amino acids demonstrated that the copepod *Metridia pacifica* synthe-

sizes coelenterazine from L-phenylalanine and L-tyrosine (see structure of coelenterazine, Figure 2.2).

It is quite likely that other species will be shown to introduce coelenterazine in the food chain, for the benefit of both bioluminescent
and nonbioluminescent organisms. Coelenterazine has indeed been
shown to have very potent antioxidant capabilities; it is, for example,
an excellent quencher of the harmful superoxide ion, O_2^-. It has been
speculated, in fact, that coelenterazine could have been selected first
for its antioxidative properties, and only later in evolution for its
light-emitting potential in the course of an oxidation reaction (see
Chapter 9).

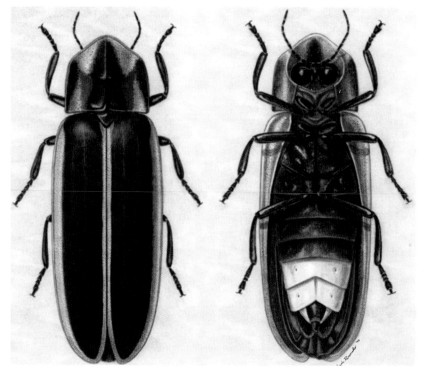

Figure 3.1. Dorsal and ventral views of a female firefly *Photinus pyralis*. The male light organ is larger.

chapter three

FIREFLIES AND OTHER BEETLES
*Luciferase-Dependent Bioluminescence Color
and Rhythmic Displays*

When we talk of bioluminescence, most of us think of fireflies such as *Photinus pyralis* (Figure 3.1). Fireflies (Lampyridae) are not flies; they are beetles (Coleoptera). Other beetles, such as the railroad worms (Phengodidae), which have both red and green lanterns, and the click beetles (Elateridae) are equally beautiful and in some ways even more intriguing.

All bioluminescent beetles—and there are thousands of species—are believed to use the *same* (beetle) luciferin (Figure 3.2) and closely related luciferases. All are also believed to emit light as the result of the same two-step reaction involving adenosine triphosphate (ATP),* even though the color of emission varies from green to red depending on species. This is one of the unique features of all beetle bioluminescence: no accessory protein, such as GFP, is used to shift the emission color. We can focus on the chemical mechanism of the firefly reaction with confidence that it is very likely to be shared by all bioluminescent beetles.

Early experiments by Dubois in 1885 had shown that after a cold-water extract of fireflies tails (assumed to contain luciferase) had stopped emitting light, mixing it with a hot extract of firefly tails (assumed to contain luciferin, but no luciferase because of the heat treatment) restored light. However, experiments in the 1940s, inspired by the then-new knowledge about ATP, showed that a bright emission could also be restored to the cold-water extract simply by

---

* ATP, adenosine triphosphate, is an "energy storage molecule" in all living cells, often called "the universal energy currency in biological systems." ADP stands for adenosine diphosphate, and AMP for adenosine monophosphate (see Glossary). The hydrolysis of 1 mole of ATP to 1 mole of ADP releases 7.3 kcal.

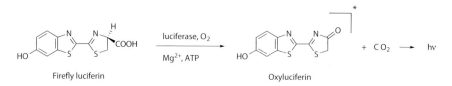

Figure 3.2. Firefly luciferin, considered to be the luciferin of all luminous beetles. The molecule is a benzothiazoyl-thiazole whose reaction with ATP and oxygen forms excited oxyluciferin, which emits light. Chemically, firefly luciferin does not resemble either cypridinid luciferin or coelenterazine, or any other known luciferin.

the addition of ATP. The end of light emission in cold-water extracts was thereby shown not to be due to the loss of luciferin, but to the loss of ATP; thus, the extract still contained both luciferin and luciferase after light emission had ceased.

No wonder, therefore, that ATP was thought at first to be the source of the energy of the green bioluminescence of fireflies. But it cannot be. The energy of a photon of yellow-green light is about seven times greater than that released in the hydrolysis of ATP to ADP. There exists no physical mechanism for the creation of a high-energy photon from seven small-energy photons.

Firefly luciferase, in fact, catalyzes two successive reactions. In the first, requiring magnesium ions (Mg^{2+}), the formation of an intermediate, a luciferyl adenylate, takes place without light emission. Then this reacts with oxygen to form an unstable cyclic peroxide intermediate, whose breakdown releases the energy for the emission of bioluminescence, as in the case of *Vargula* and *Renilla* (cypridinids and coelenterates). Once again, as in all bioluminescence, the role of oxygen and peroxides is of unique importance (see Box 3.1).

Early on it was noticed that many factors can drastically alter the color of the light *in vitro*, without involving any accessory protein, such as GFP. Some firefly luciferases, such as that of *Photinus pyralis*, are sensitive to pH. At an alkaline pH (~8.5), the emission peaks in the yellow at 560 nm, while around an acidic pH (~6.4), the peak falls in the red at 620 nm and the intensity is much lower (Figure 3.3). This suggested the possibility of two different emitters. Indeed, the chemical structure of oxyluciferin seemed compatible with that in-

Box 3.1. The bioluminescence reaction of fireflies.

As in the case of *Vargula* (Box 1.2), painstaking experiments with firefly luciferase using isotopically labeled O_2 have also established the role of a dioxetanone intermediate as the source of the energy released by the emission photons in the firefly luciferase reaction. But the overall reaction is more complex. When bound to luciferase, firefly luciferin first reacts with ATP in the presence of Mg^{2+} to give a luciferin adenylate intermediate. This adenylate reacts with oxygen to make a presumed dioxetanone (never isolated), which breaks down to generate CO_2, AMP, and oxyluciferin in its excited state, then light emission.

The first of the two successive reactions, the adenylation, is slow; the second is very fast. The adenylate intermediate can be synthesized chemically and mixed with luciferase and oxygen in the absence of ATP; it takes only about 60 msec from the time of mixing the synthetic adenylate and oxygen to the peak of light emission.

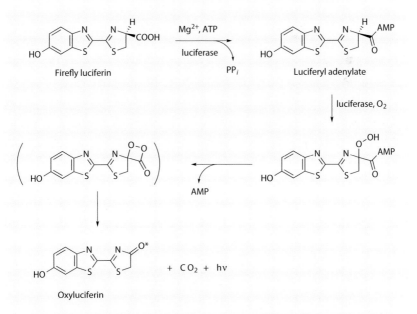

Steps and intermediates in the firefly luciferase reaction. The reaction with ATP forms the intermediate luciferyl adenylate, which reacts with oxygen, forming a cyclic peroxide, whose breakdown forms oxyluciferin in the excited state.

terpretation: one emitter could be in the keto form of oxyluciferin, as shown in the figure in Box 3.1, the other in its enol form (Figure 3.4, left), with the keto emitting in the red and the enol in the green.

For a number of years, this was the accepted explanation, never mind that the red-emitting keto would be formed first and that a significant amount of additional energy would be needed to make the green-emitting enol. This energetic difficulty appears not to have ever been discussed in the early years. In any case, a dimethyl-substituted luciferin analog (Figure 3.4, right), which cannot form the enol, has now been shown to emit green light, a fatal blow to the keto/enol interpretation. Moreover, as Figure 3.3 clearly shows, the green and red emitters are not in equilibrium; only the green emitter is pH sensitive. If the two emitters were in equilibrium, the emission would be more intense at pH 6.4 than at pH 8.5 in the red part of the spectrum.

It was also observed, early on, that mixing firefly luciferin and ATP *in vitro* with luciferases from different firefly species resulted in different emission spectra, even though the luciferin and pH were the same in all cases. This established that the determinant of the color of emission is the enzyme itself. *Different beetles emit different*

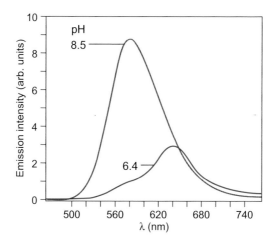

Figure 3.3. The effect of pH on the spectrum and intensity of *Photinus pyralis* emission. Note that the intensity of the emission in the red (e.g., from 620 to 740 nm) is unaffected by pH.

Figure 3.4. Firefly oxyluciferin in enol form (left) and dimethyloxyluciferin (right).

colors of light because they have slightly different luciferases, not because they have different luciferins or accessory fluorescent proteins. Beetle luciferases are the only luciferases known to have a role in color determination *in vivo.*

Many beetle luciferases have now been cloned, sequenced, and expressed in *E. coli* and in eukaryotes. All are proteins of about 60 kDa, ~550 amino acids long; experimental substitutions of one or several amino acids, here and there in the sequence, have been found to alter the color of emission, without giving straightforward clues as to what is responsible for the spectrum.

Critical to the goal of understanding the factors responsible for emission color was to establish the three-dimensional structure of the enzyme. The crystal structure of *Photinus* luciferase was the first of the luciferases to be obtained, in 1996. It shows a two-domain protein, with a large N-terminal domain and a much smaller C-terminal domain linked by a short loop, with a cleft accessible to water between the two domains (Figure 3.5). This cleft is close to the location of the enzyme active site, where oxygen reacts with the luciferyl adenylate; many amino acids critical for emission color are known to be located in this region. The first step of the reaction may cause the cleft to close, excluding water and creating a hydrophobic site for the oxidation of the adenylate.

A recent spectroscopic study considers the luciferase at the active site (that is, at the location of the luciferin within the enzyme) as if it were a solvent. It concludes that two factors determine the color of the emission: the polarity around the OH- group on the benzothiazole of excited luciferin, and the presence of basic amino acids susceptible to covalently bind to this OH group. It is well know that the acidity or alkalinity of a solution can change the color of a dye—pH test strips come to mind. By the same token, the environment around

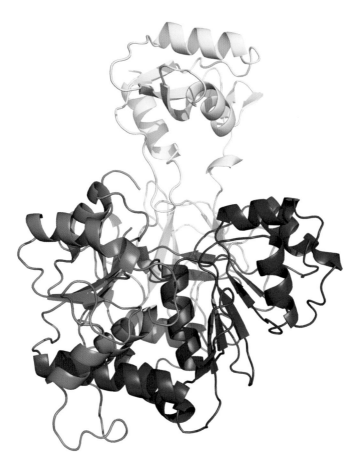

Figure 3.5. Ribbon representation of firefly luciferase. The C-terminal domain (yellow, *at top*) is connected by a short flexible loop to the larger N-terminal domain (blue, purple, and green subdomains). The binding sites for both ATP and luciferin are located on the N-terminal domain. The active site appears to be located on the N-terminal domain near the cleft between the two domains.

excited luciferin within the enzyme may change the color of the emission. We may still be a long way from being able to explain how substituting a single histidine by a tryptophan in the luciferase, for example, may shift the emission from green to red, as it does; but it is fair to say that we now know that the reaction generates a single emitter that carries a hydroxyl group extremely sensitive to its environment. It is remarkable that firefly luciferase is the only known luciferase to so sensitively determine the color of emission.

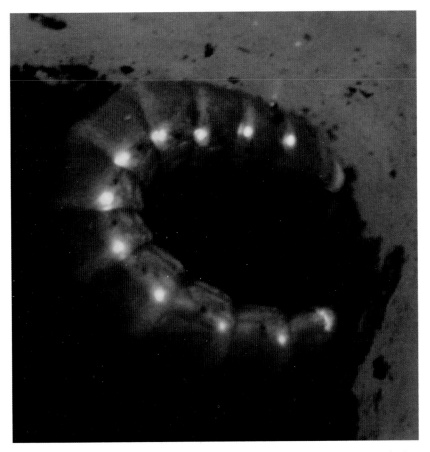

Figure 3.6. The railroad worm *Phrixothrix hirtus;* its luciferase is similar to that of other beetles and its luciferin the same as that of *Photinus pyralis.*

Explaining emission color has always been especially challenging when confronted with *Phrixothrix hirtus,* the famous phengodid railroad worm of South America. The stunning larva of this beetle has green lanterns along its body and red lanterns on the head (Figure 3.6). Yet both light organs use the same luciferin; it is their luciferases that are different. Thus, *in vitro* red and green emissions are obtained with the same luciferin but with luciferases extracted from two regions of the larva: the head and the body. We can therefore assume that this is the case also in the animal.

The chemistry of light emission is only one intriguing side of firefly bioluminescence. Another is the kinetics and control of flashing. How are these light pulses, precise to better than 0.2 sec, turned on, and then off? Although this is still debated, the evidence indicates that oxygen entry triggers the flash onset by reacting rapidly with an enzyme intermediate accumulated in the absence of oxygen. The decay of light emission, on the other hand, does not require the removal of oxygen, at least *in vitro*, because the flash itself represents the rapid reaction of the accumulated enzyme intermediate, which is reformed only slowly. This means that the light can decay to a near zero level without the need for oxygen removal.

The light organ consists of many photocytes organized in rosettes, stacked around the trachea and tracheoles, which bring in oxygen. Within these photocytes, small organelles, called peroxisomes, contain luciferase, adenylated luciferin, and ATP; the peroxisomes are maintained anaerobic between flashes. Nerve impulses from the brain trigger the flashes, terminating not on the photocytes but on tracheal end cells (TEC) that surround the tracheoles that lead to photocytes (Figure 3.7).

Exactly how the access of oxygen to photocytes is regulated is debated, but the fact that the photocytes are not innervated directly is key. One theory is that entry of air *via* the tracheoles is normally blocked by fluid and that the nerve impulse somehow rapidly removes the fluid and allows air entry. A more recent proposal is that oxygen entry is not blocked and that peroxisome anaerobicity results from vigorous mitochondrial respiration (which would maintain a ready supply of ATP). The flash is proposed to occur when nitric oxide (NO), released by the nerve impulse, enters the photocyte, and immediately blocks mitochondrial respiration, thus allowing a spike in oxygen concentration, hence a flash. In this view, it is the activation of nitric oxide synthase (NOS) that controls the flashes. Other hypotheses have been proposed, but there is no consensus on the mechanism.

It is well established that flashing is used for communication in courtship, but there are many different species-specific patterns. In some American species, the flying male flashes in search of a female,

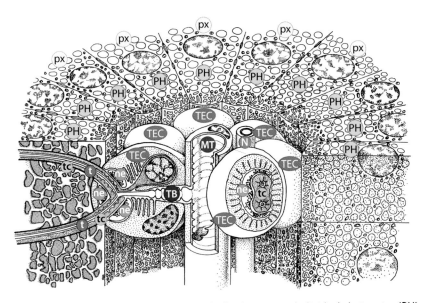

Figure 3.7. The rosette organization of the firefly photocytes. Individual photocytes (PH) with luciferase-containing peroxisomes (px) are arrayed like pie segments surrounding the main trachea (MT). Tracheoles (t), formed from tracheolar cells (tc), bring oxygen to photocytes; they penetrate laterally via horizontal branches (TB), which are surrounded by tracheal end cells (TEC). These are innervated by nerve endings (ne), branches of the central nerve (N).

who responds with a flash after a species-specific time interval; they then engage in a courtship that involves several confirmatory exchanges between the two parties, as the male approaches his mate (Figure 3.8).

But a most extraordinary example of firefly courtship is found along some rivers of Southeast Asia, from the Ganges, east of New Delhi, to Thailand, Singapore, the Philippines, Papua New Guinea, and New Britain. Thousands of male fireflies of the genus *Pteroptyx* perch on tree leaves at night and flash in absolute unison, presumably to attract the attention of females (Figure 3.9). Each flash lasts only a tenth of a second, and the interval between flashes is roughly 1 second, depending on species. Amazingly, the synchronization is exact to 20 msec between groups of fireflies in distant parts of a tree, and even from tree to tree along a river. The females also flash and

Figure 3.8. A female firefly perched on a twig, flashing in response to a courtship inquiry from a male.

glow, but more modestly and not in synchrony; they seem to stay among groups of males, and probably pick and choose the best displaying candidate.

But why go to Southeast Asia when you can see synchronous fireflies in our own Great Smoky Mountains National Park? The town of Elkmont, Tennessee, is now attracting crowds to the no-longer-well-kept secret of thousands of synchronous *Photinus carolinus* doing what the locals call their annual light show during two days in June. Laboratory experiments with *P. carolinus* suggest that females respond most favorably to males synchronized to their own flashing frequency.

Lest you think of female fireflies as utterly passive, consider the case of the so-called femmes fatales among American fireflies. Females of the genus *Photuris*, known to be carnivorous, mimic the flashing code

Figure 3.9. Synchronous flashing fireflies in a tree in Southeast Asia. Dotted tracks are from individual fireflies flashing regularly during flight as they move from one leaf to another. Double exposure, the first before dark that outlines the tree; the second was a time exposure (perhaps 30 sec) after dark, showing the fireflies.

Figure 3.10. Scores of termite mounds in the *cerrados* of central Brazil populated by click-beetle larvae (*Pyrearinus termitilluminans*) living in networks of tunnels opening to the outside, where the beetle larvae perch.

of the females of another genus, *Photinus,* to attract *Photinus* males, and then devour them. This meal provides them not only with nutrients but also with a specific steroidal compound unpalatable to many predators such as spiders (unattractively named lucibufagin), which *Photinus* can produce but *Photuris* cannot. Female *Photuris* transfer this lucibufagin, and therefore protection, to their eggs and offspring.

Far from the trees of East Asia, other species of bioluminescent beetles co-inhabit termite mounds 6 to 7 feet in height in the *cerrados* of central Brazil (Figure 3.10), creating sites of astonishing beauty. Starting right after sunset, scores of mounds can be seen to be illuminated with hundreds of points of green light (Figure 3.11), emitted by click-beetle larvae (*Pyrearinus termitilluminans*), which live in networks of tunnels opening to the outside. Sitting on their balconies, they catch and eat all the flying prey their green thorax lights attract, ants and their termite hosts among them. In turn, this teem-

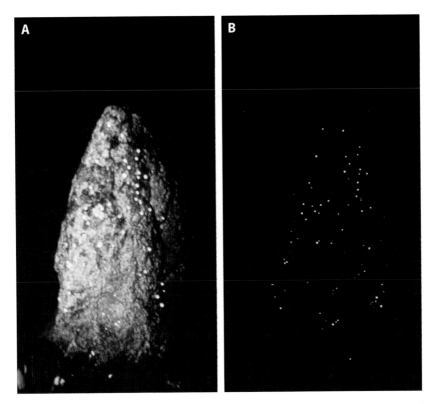

Figure 3.11. A. A double exposure image at night, the first taken by flash and the second in the dark, showing dozens of points of green light from the larvae. B. Image from a photograph taken in the dark.

ing community of life attracts party crashers, from birds and bats to frogs, spiders, scorpions, and plants. If it's easy to see what the bioluminescent beetles gain from co-habiting with termites—bed and breakfast—the opposite is not evident.

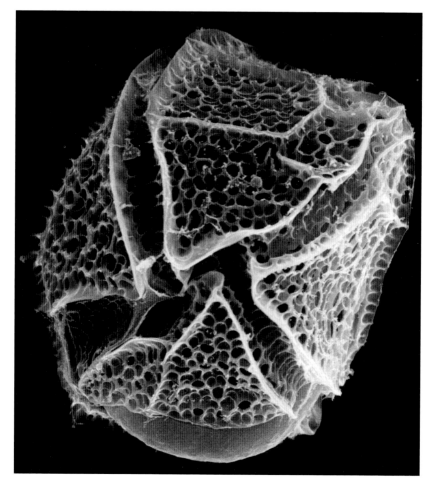

Figure 4.1. A *Lingulodinium polyedrum* cell (diameter, ~40 µm) showing the prominent groove, called the girdle, and the highly sculptured cell wall divided into cellulose plates. Each plate is enclosed in a vesicle; the two flagella (*not seen*) emerge from a single point in the cell wall.

chapter four

DINOFLAGELLATES AND KRILL

The Sparkling Clocks of the Oceans and Bioluminescent Shrimp

The small unicellular dinoflagellates, which occur ubiquitously in the oceans, are responsible for the beautiful flashes that occur when seawater is disturbed at night, the bioluminescence of the ocean. The chemistry of their bioluminescence is altogether different from the other major systems. But here again, as in the cases of the small crustacean of Chapter 1 and of the jellyfish of Chapter 2, a bioluminescence system similar to that of the dinoflagellates is also found in a phylogenetically remote group, shrimps known collectively as krill.

The Dinoflagellates

Many of us have seen, at one time or another, a sea sparkling with flashes of light in the wake of a boat, on our oars or in breaking waves (Figure I.1). These tiny light sources are bioluminescent single-cell algae, called dinoflagellates. They comprise thousands of species, the majority in the oceans (with many luminescent ones), but many freshwater species also, although none of those are luminescent. The "dino" part of the name comes from the Greek *deinos*, which means whirling; the "flagellate" part, from Latin, is obvious.

The particular bioluminescent dinoflagellate we know most about, *Lingulodinium polyedrum* (formerly called *Gonyaulax polyedra*), usually responsible for red tides and bioluminescence in the ocean along the coast of Southern California, is about 40 μm in diameter (Figure 4.1). It has rigid cellular plates referred to as armor, and two flagella, one circling its very impressive girdle and the other at right angles.

It is photosynthetic, as many dinoflagellates are. Upon mechanical stimulation, it emits flashes of blue light peaking at 470 nm. Why do they emit? One idea is that the flashes deter small assailants directly; another, called the burglar alarm theory, is that the flashes reveal the presence of small predators, the true nemesis of the dinoflagellates, to bigger ones, such as crustaceans that feed on the small ones.

In *L. polyedrum*, the bioluminescence is emitted by small organelles, ~0.4 µm in diameter, several hundred per cell; these are called "scintillons," the flashing units (Figure 4.2). Immunolabeling experiments and electron microscopy have shown that the scintillons are located peripherally in the cytoplasm (Figures 4.3 and 4.4); they hang like little drops into the acidic vacuole, surrounded by the vacuolar membrane (Figure 4.5). Upon disruption of the cells at pH 8, the scintillons break off as small vesicles, with necks resealed, and can be isolated with bioluminescence activity; they can be made to emit a flash of light simply by acidifying the medium to pH 6. Scintillons contain mainly luciferin and two densely packed proteins, one the luciferase and the other a luciferin-binding protein (LBP), which has no catalytic activity.

As in other cases, the terms "luciferase" and "luciferin" are generic, related only functionally, not structurally, to those of any other bioluminescent organism. Also as in other cases, the luciferases of all dinoflagellates studied so far are able to catalyze the light-emitting oxidation of not only their own luciferin but also that of other dinoflagellates, such as the photosynthetic species *Pyrocystis lunula* (Figure 4.6), from which the luciferin was purified and its structure determined.

The chemical structure of this luciferin, which is likely to be the luciferin of all dinoflagellates and is therefore referred to here as "dinoflagellate luciferin," has been established to be an open-chain tetrapyrrole (Figure 4.7), clearly related to chlorophyll. This is perhaps not surprising, since it functions in a photosynthetic organism. (Note, again, that this luciferin bears absolutely no chemical similarity to cypridinid luciferin or to coelenterazine, nor to bacterial or

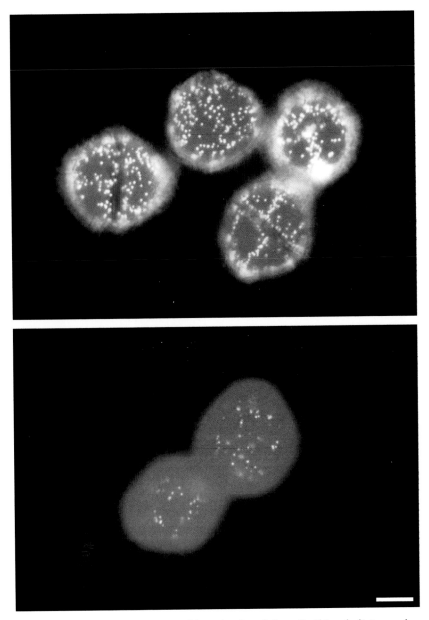

Figure 4.2. Fluorescence images of live dinoflagellate cells (*Lingulodinium polyedrum*) showing many luminous organelles (scintillons) at night (top) and very few during day phase (bottom). The scintillon fluorescence is from the luciferin; the red background is the fluorescence of the abundant chlorophyll. The scale bar is 15μm.

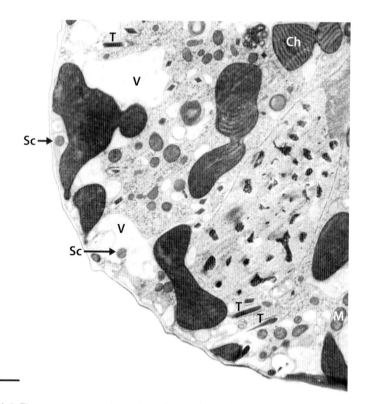

Figure 4.3. Electron microscopic section of a quadrant of a *L. polyedrum* cell; two scintillons (Sc) are indicated by arrows; vacuole (V); chloroplast (Ch); mitochondria (M); trichocyst (T). Scale bar, 1μm.

insect luciferins). Although the chemical structure of the product of the luciferase-catalyzed reaction of the luciferin with oxygen has also been established, the precise chemical mechanism of the reaction is still unknown—for example, we don't know if it involves a peroxidic intermediate, although we hypothesize that it does.

Curiously, the bioluminescence emission spectrum matches perfectly the fluorescence of the *unreacted* luciferin, while the product of the reaction is not fluorescent *in vitro*. This paradox pinpoints our ignorance; the emitter may be an unstable fluorescent intermediate, or an intermediate that is fluorescent only when protein bound, or, indeed, an unreacted luciferin molecule by energy transfer from the excited oxidized product.

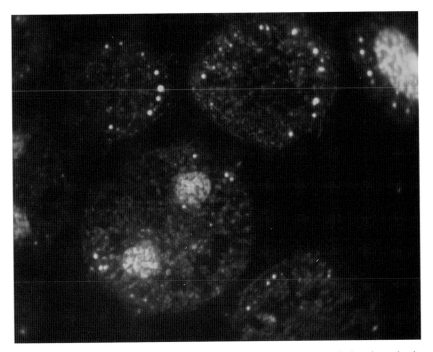

Figure 4.4. Scintillons of *L. polyedrum* cell from night phase are labeled with antibody against LBP showing their peripheral location. The nucleus is also labeled, but the nuclear protein responsible for labeling is not known.

In contrast, a great deal is now known about the two proteins involved in the light-emitting reaction, the luciferase and the luciferin-binding protein (LBP). This luciferase reaction is unusual; the catalysis of the oxidation of luciferin by O_2 is sharply dependent on pH. In alkaline solution, the luciferase has no activity, but it gains activity when the solution is made acidic (pH 6). Remarkably, LBP activity mirrors this; it binds the luciferin at high pH and releases it when the solution is acidified, allowing it to bind in turn to luciferase and be oxidized with light emission. These pH dependencies of luciferin sequestration and luciferase activity result in a reaction that is truly triggered by a pH change, not simply dependent upon it, as enzymatic reactions are.

How does this triggering occur? Given the structural disposition of scintillons (Figure 4.5), and the fact that mechanical stimulation

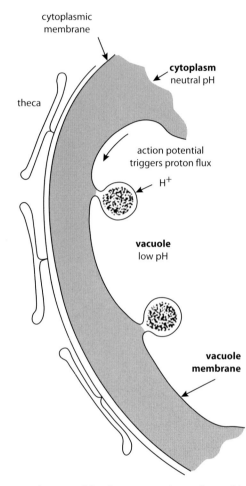

Figure 4.5. Schematic diagram of the disposition of scintillons of *Lingulodinium poly-edrum*. They are represented as droplet-like organelles, formed as cytoplasmic out-pockets hanging in the vacuole, surrounded by the vacuolar membrane.

of the cell initiates an action potential in the vacuolar membrane, it was hypothesized that there are channels in the scintillon membranes specific for hydrogen ions, and that these are opened upon passage of the action potential, allowing the ions to enter and acidify the scintillons. This activates the luciferase and LBP, triggering the flash. A gene for such a channel has recently been found in a nonluminous dinoflagellate and shown to have the specific properties required to function as hypothesized.

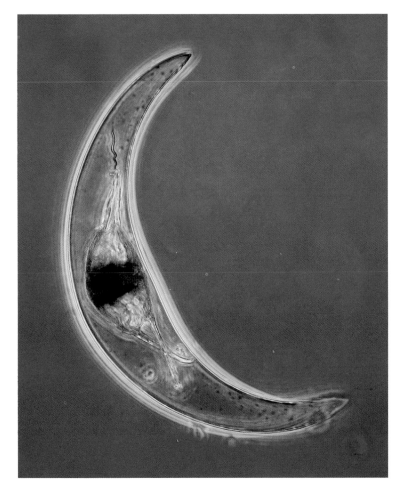

Figure 4.6. A *Pyrocystis lunula* cell by light microscopy.

Structurally, the luciferase protein is also very unusual; it is a chain of three contiguous domains with very similar amino acid sequences, each 377 amino acids long, preceded by an N-terminal sequence (Figure 4.8). Each of these domains alone is catalytically active, and each domain also shows the same pH dependency as the full-length enzyme. Thus, there are two sets of redundancies in this enzymatic reaction: two proteins for regulating the pH dependency of the reaction rate, and three active sites in a single molecule of luciferase.

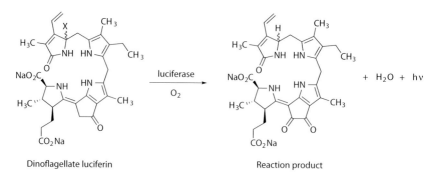

Dinoflagellate luciferin   Reaction product

Figure 4.7. The chemical structure of *Pyrocystis lunula* luciferin (an open chain tetra-pyrrole, considered to be the luciferin of all dinoflagellates) and the product of its luciferase-catalyzed oxidation. The similar chlorophyll molecule (*not shown*) is a closed tetrapyrrole ring.

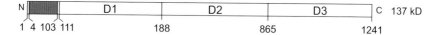

Figure 4.8. Within a single protein molecule, the luciferase of *L. polyedrum* comprises three homologous domains, D1, D2, D3, each with an active site. These three domains are preceded by an N-terminal sequence of about 100 amino acids of unknown function.

The 3D geometry of the full-length luciferase is not yet solved, but the crystal structure of one of the catalytic domains is known and goes a long way to explaining the pH effect. It suggests that the most likely position of luciferin, when bound to the enzyme, is within a barrel-shaped structure in the lower region of the structure, as shown in Figure 4.9. Access to this site may be had only through a channel formed by the helices located at the top of the figure, and at pH 8 this channel may be too small to allow entry of luciferin. But there are four key histidines located in the channel region; they have different structures at pH 6 and pH 8, allowing or not the luciferin access to its binding site on the luciferase, depending on pH.

The bioluminescence of *L. polyedrum* is of particular interest also because it is rigorously controlled by the circadian clock (the daily

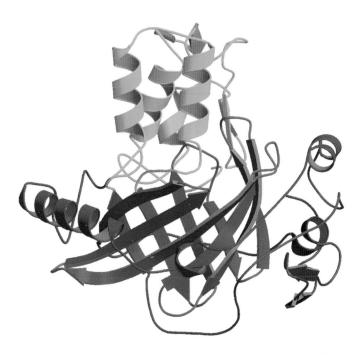

Figure 4.9. A ribbon diagram of the crystal structure of domain 3 of the *L. polyedrum* luciferase. The three helices at the top form the entry channel for luciferin, which may be either open or closed depending on pH. Luciferin binds to the enzyme in a pocket formed by the strands in the lower part of the figure.

rhythm) of this single-cell organism. Shaking a flask of cultured dinoflagellates during daytime produces only a very weak emission, but doing so at night produces a spectacular fireworks display. Kept in constant conditions in the laboratory (dim light, since it is a photosynthetic organism) after having been exposed to daily light and dark cycles, the cells continue to be rhythmic, emitting ten to a hundred times more bioluminescence during subjective night phases, that is, when it would have been night. They do so with a phenomenally accurate rhythm (Figure 4.10) that has made this dinoflagellate a star of clock research.

Luciferase, luciferin, the binding protein, and, in fact, the scintillons themselves are all destroyed at the end of each night, only to be synthesized anew at the beginning of the next night. Therefore, cells used to isolate the scintillons and proteins must be in their night

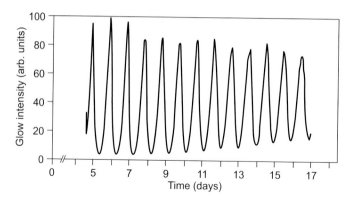

Figure 4.10. The circadian rhythm of the steady spontaneous glow of bioluminescence of *Lingulodinium polyedrum*. Glow intensity in arbitrary units.

phase (Figure 4.2, top); in the day phase, there remain fewer than 10 percent of the night phase number of scintillons (Figure 4.2, bottom). Another equally accurate circadian rhythm regulates the daily up-and-down migration of the cells in the water column. During daytime they remain near the surface, where they catch the sunlight, while at night they migrate downward, where inorganic nutrients are more available.

Pyrocystis lunula (Figure 4.6), the species to which we owe our knowledge of the structure of dinoflagellate luciferin (Figure 4.7), uses a luciferase very similar to that of *L. polyedrum*, but it appears to lack a luciferin-binding protein (LBP). How the role of LBP is carried out in this case is not known. And while *Pyrocystis* bioluminescence is also circadian, it achieves this rhythmicity in a totally different and more conservative way—not by a daily destruction of scintillons, but by moving them to the periphery of the cell at night from positions near the nucleus during the day (Figure 4.11). A more detailed understanding of how light emission is thereby controlled is also lacking.

Among the heterotrophic (nonphotosynthetic) bioluminescent species, *Noctiluca scintillans* is up to 1 mm in diameter and a formidable predator, though unarmored. It is equipped with a thick tentacle, a modified flagellum, which serves to capture prey and push it down its gullet (Figure 4.12).

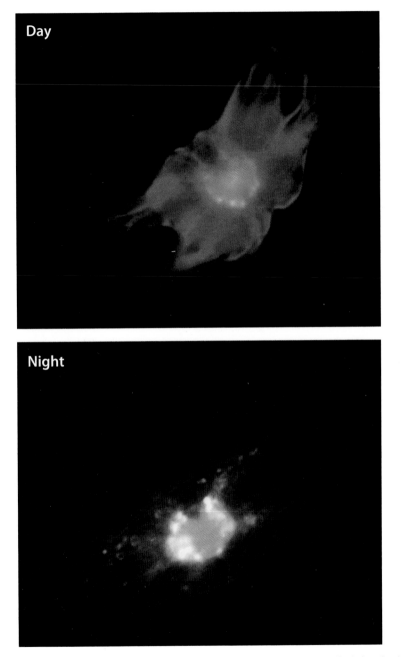

Figure 4.11. *Top*. Day-phase *P. lunula* cell with scintillons located centrally (white bodies) and chloroplasts expanded and located closer to cell surface. *Bottom*. Night-phase cell with scintillons (small dots) located peripherally and chloroplasts (red fluorescence) centrally. Larger green bodies in center are organelles called PAS bodies. In both images the outline of the moon-shaped cell wall are only scarcely visible.

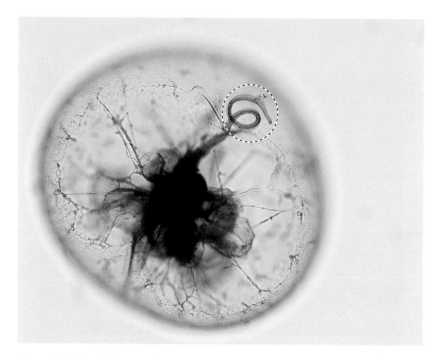

Figure 4.12. *Noctiluca scintillans*. A heterotrophic dinoflagellate, it ingests food with the aid of the peducle (*circled*).

Its bioluminescent system and luciferase gene reveal some important similarities with its photosynthetic cousins, and also some remarkable differences. It was found that only one protein, not two, is involved in the light-emitting reaction of *Noctiluca*. In this protein, a segment of the *L. polyedrum* luciferase is fused with the entire binding protein (LBP) of *L. polyedrum* to make the single *Noctiluca* luciferase. Also, this two-part protein of the nonphotosynthetic *Noctiluca* catalyzes the *in vitro* bioluminescence reaction with *Pyrocystis* luciferin, which is so clearly a derivative of chlorophyll. Does *Noctiluca*, which has no chlorophyll, get its luciferin from its diet, perhaps by eating its smaller photosynthetic cousins, or does it synthesize its own luciferin? A recent study of a different nonphotosynthetic species indicates the latter.

Bioluminescent Shrimp

Speaking of diet, euphausiid crustaceans known as krill are thought to acquire the elements of their bioluminescent system by ingestion because their luciferase and luciferin cross-react with those of dinoflagellates. Krill collectively comprise dozens of species, among them the Antarctic krill, *Euphausia superba* (Figure 4.13). Two to three inches long, these crustaceans emit bursts of blue bioluminescence from pairs of pivoting light organs along their bodies, each equipped with a lens and a reflector, as well as from another pair of light organs on their eyestalks. They assemble in huge swarms of

Figure 4.13. The bioluminescent krill, *Euphausia norvegica*, about 4 cm in length, photographed in room light (top), and the same animal photographed by its own light from the ventral side (bottom).

tens of thousands per cubic meter everywhere around the Antarctic, possibly using their bioluminescence to stay in close touch with the gang at night; of course, no one knows for sure. Krill filter-feed on plankton, a lot of it photosynthetic, and are themselves the food of many animals, from ctenophores to fishes, whales to sea birds. It is estimated that crabeater seals eat millions of tons of krill in a year, while a single female *E. superba* lays 10,000 eggs several times over the same time span! Here we are definitely not talking of a rare or unimportant curiosity of nature, but of a major part of the global food chain.

The bioluminescence system of *E. superba* has not been studied, but what is known of the bioluminescence of other krill (*Meganyctiphanes norvegica* and *Euphausia pacifica*) is intriguing and puzzling. Two components have been isolated that in the presence of oxygen react with emission of blue light: a luciferin, again a fluorescent tetrapyrrole, apparently very unstable, and a luciferase, a very large protein (molecular mass, about 600,000 kDa). The luciferin structure, as well as the bioluminescence emission spectrum, is similar to those of dinoflagellates, but the luciferase appears quite different. Enzymes and substrates of dinoflagellates and euphausiids were shown, already 30 years ago, to cross-react with emission of light, with the large krill luciferase being the more accommodating of the two enzymes; it gives brighter light with dinoflagellate luciferin than dinoflagellate luciferase gives with krill luciferin. However, the structure and amino acid sequence of krill luciferase have not been determined, so its possible relationship to dinoflagellate luciferase is not known.

However, again the question arises: do the krill synthesize their own luciferin or acquire it nutritionally? Chemistry appears to favor the first alternative; two of the four pyrrole groups (A and B) carry different substituents in krill and dinoflagellate luciferins, while the "business parts" of the two luciferins (groups C and D) are identical. So if the krill luciferin is transferred from dinoflagellates, it is modified before use.

One last note: As in the case of the *Pyrocystis* luciferin-luciferase reaction, the bioluminescence spectrum of *E. pacifica* matches the fluorescence spectrum of the unreacted luciferin, and the enzymatic oxidation product *in vitro* is not fluorescent, possibly because it is not enzyme bound.

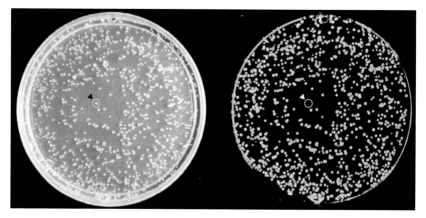

Figure 5.1. Colonies of luminous bacteria photographed in room light (left) and by their own light (right). Note the marked colony in center of the plate at left; as seen at the right, there is no light at that location. Such dark variants may be frequently encountered but can revert to the bright form (see text).

chapter five

BACTERIA

Bacterial "Communication," Symbioses, and Milky Seas

If many fishes have co-opted the light-emitting systems of cypridinids and coelenterates, many other ocean denizens, as well as notable terrestrial species, acquire their bioluminescence by hosting luminescent bacteria. In doing so, they acquire a complete luminous system—luciferin, luciferase, and all the trimmings—and enter into a two-way, very specific symbiotic relationship between hosts and guests, with costs and benefits to both.

Luminescent bacteria (Figure 5.1) are about a thousand times smaller and also simpler than *Vargula* and cells of other eukaryotes. Bacterial DNA is in a tangle, not in a nucleus surrounded by a membrane, and not organized in chromosomes. Yet exploring the bioluminescence of bacteria has taught us lessons that bear on many aspects of life, including our own health.

The biochemistry of the bacterial luciferase reaction, shared by all species, such as *Vibrio fischeri*, follows a now-well-described complex mechanism. The role of the luciferin is played by a fluorescent molecule, reduced riboflavin phosphate, also called flavin mononucleotide ($FMNH_2$), which is a participant in the respiratory chain of all aerobic cells. Bacterial luciferase is not a simple enzyme either, as it catalyzes the oxidation of not one but two substrates, the flavin and a long chain aldehyde (RCHO). The overall reaction, with all intermediates bound to the enzyme, is

$$FMNH_2 + O_2 + RCHO \xrightarrow{\text{luciferase}} FMN + RCOOH + H_2O + h\nu$$

where RCOOH stands for a long-chain acid and FMN for the oxidized flavin. What becomes the emitter is a hydroxyl derivative of the isoalloxazine part of the flavin molecule, FMNH-OH (see Box 5.1 for the presently accepted, more detailed reaction mechanism.)

In contrast to the systems discussed in the preceding chapters, where chemists zeroed in on a cyclic peroxide as the penultimate intermediate leading to the excited state, the bacterial reaction forms a linear peroxide in the first step. This has been isolated, but the intermediate penultimate to the excited-state formation has eluded identification.

Box 5.1. The bioluminescence reaction of bacteria.

Bacterial luciferase catalyzes the oxidation of not one but two substrates, the flavin and the aldehyde. Although there are significant differences between the luciferases of the different species of luminescent bacteria, all are composed of two homologous subunits (α and β, of ~40 and ~35 kDa), with the active center located on the α subunit. The presumed emitter, the transient hydroxyflavin, is *not* the final product, which is FMN. *In vitro*, aldehyde chain lengths from about 6 to 18 result in light emission; *in vivo*, the natural aldehyde has been identified as tetradecanal, $CH_3\text{-}(CH_2)_{12}\text{-}CHO$.

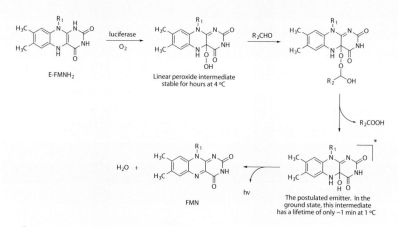

Substrates, intermediates, and products in the bacterial luciferase reaction; all intermediates are enzyme-bound.

In some species or strains the color of the light emitted by living cells is significantly blue- or red- shifted, even though the reaction with isolated luciferases still peaks in the blue at ~490 nm. In *V. fischeri* strain Y-1, for example, the emission *in vivo* is yellow (λmax ~540 nm; Figures 5.2 and 5.3, left), but that of the isolated luciferase reaction is blue (Figure 5.3, right); this is attributed to a second protein, an "antenna" protein called yellow fluorescent protein (YFP) that has its own flavin fluorophore. Purified YFP from *V. fischeri* added to a reaction with purified luciferase results in the emission of mostly yellow light (Figure 5.3, right), as in the strain Y-1 cells (Figure 5.2).

This color shift is reminiscent of the GFPs in coelenterates, where energy transfer from excited coelenteramide to GFP has been suggested. But evidence in the bacterial yellow-emitting system indicates that its "antenna" protein appears to actually enter into the bacterial reaction, detouring it toward formation of the excited state of the antenna protein, and hence a different emission color. The functional importance for such spectral shifts has not been elucidated; however, strains with a blue-shifted emission (with a different antenna protein) occur primarily at depths of ~600 m in the ocean.

Years of work on the enzymology and biochemistry of bacterial bioluminescence paid off handsomely when gene sequence data led to a wider understanding of a surprising observation: when luminous bacteria are inoculated at a low density in a fresh liquid medium with all needed nutrients, the population starts immediately to grow exponentially, as measured by the increase in the number of cells (Figure 5.4A). However, for a while (hours) there is no increase in light emission. In fact, if light intensity is measured as a function of time, it starts increasing only when the density of cells reaches a certain level (about 100 million cells per milliliter; the open ocean typically has 1 cell per milliliter), after which the light intensity increases much faster than cell growth.

The critical finding was that when the same experiment was repeated in a "conditioned" medium (obtained from a brightly emitting culture after cells were removed), the development of luminescence was not delayed. The correct interpretation of this simple experiment was a breakthrough. The bacteria produce an "autoinducer," a small

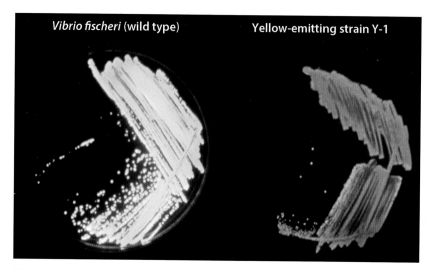

Figure 5.2. Cultures of blue-emitting wild type *V. fischeri* (left) and yellow emitting *V. fischeri* strain Y-1 (right).

molecule that diffuses freely in and out of the cells, which upon reaching a critical level in the medium causes the genes involved in the synthesis of the luciferase to be activated and transcribed.

Over the years, thanks to the work of many researchers, the picture is now clear and detailed. First, the autoinducer involved in the particular bacterial species of the experiment described above was found to be a homoserine lactone (Figure 5.4B), a relatively small molecule (about the size of a sugar). Second, we now understand *how* it regulates the expression of bioluminescence.

It all takes place on the rather small piece of DNA, about 7000 base pairs, which encodes all the necessary parts of the bioluminescent system. This "regulon" is made of two "operons" (Figure 5.5), transcribed in opposite directions. The rightward operon is responsible for the synthesis of the autoinducer synthase (I), the three enzymes responsible for the synthesis of aldehyde from the corresponding fatty acid (C, D, E), and the two subunits of the luciferase (A, B); the leftward operon produces a regulatory protein called R.

At low cell density, the autoinducer accumulates only slowly in the medium. But when its concentration reaches a certain level, it binds with protein R, which in a feedback loop activates the transcription

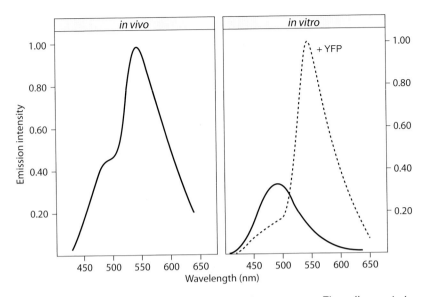

Figure 5.3. Spectral measurements of bacterial bioluminescence. The yellow emission of living cells of *V. fischeri* strain Y-1 showing shoulder in the blue (left). The *in vitro* reaction with purified luciferase (*solid line*) and (*dotted line*) luciferase with added purified YFP, both isolated from *V. fischeri* strain Y-1 (right).

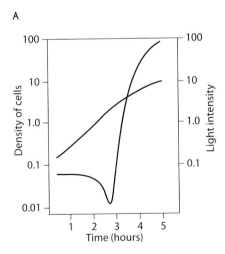

Figure 5.4. *A.* Experiment showing that in a newly inoculated bacterial culture, the cell density (left *y*-axis) starts increasing immediately and continues growing exponentially with time, whereas the bioluminescence emission (right *y*-axis) remains constant at first, then decreases before starting to increase at a rate much faster than cell density. *B.* The *V. fischeri* autoinducer (a homoserine lactone); when it reaches a critical concentration in the medium, luciferase and other related proteins are synthesized.

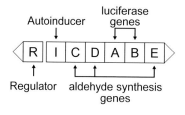

Figure 5.5. The luciferase (*lux*) operon (a cluster of genes in the DNA that are regulated together), designating the arrangement of genes responsible for bioluminescence in the bacterium *V. fischeri*. The R and I genes are located elsewhere in the chromosome in some other species.

in the rightward operon, causing the synthesis of aldehyde and luciferase, hence light emission. The entire regulon of one bacterial species has been put in *E. coli* cells, making these cells glow and be regulated exactly as if they were genuine bioluminescent bacteria.

These results clearly show that bacteria communicate chemically among themselves. If the benefits of bioluminescence studies needed advertising, here is where to start. Measuring light is easy, and its intensity can be followed as a function of time and location; in a petri dish, the one bacterial colony that glows can be picked up among hundreds that do not. This may be why the role of a secreted "activator," inferred earlier in the case of nonluminescent bacteria (*Streptococcus pneumonia*), was slower to attract attention; the phenomenon was the same, but it did not hit one in the eye, literally, as the bioluminescence of the cells did.

In the 40 years since autoinduction was discovered in bioluminescence, and the 25 years since the genes involved were sequenced and expressed in *E. coli*, the generality of the phenomenon has been broadly established; the process of bacterial communication is now dubbed "quorum sensing," an evocative term suggested by a lawyer who was told of the phenomenon. But one should not be misled into the idea that the bacteria literally count the numbers in their immediate environment; they simply respond to the inducer concentration, which will become high when there are many bacteria confined in a limited space. As was shown with "conditioned" medium, cells will synthesize the luciferase system when the medium has a high enough concentration of the inducer; adding synthetic inducer will have the same effect.

Quorum sensing is now a tremendously active field of research. Different autoinducers have been discovered, as well as cases where two or more autoinducers are at work. Among the so-called Gram-negative bacteria (so-classified on the basis of their cell walls), which include all bioluminescent bacteria, all use homoserine lactones as autoinducers, albeit different ones, specific for each bacterial species. Gram-positive bacteria use small, specific peptides to pass on information.

One life-shortening disease where quorum sensing is suspected to play an important role is cystic fibrosis, caused by the bacterium *Pseudomonas aeruginosa*. Like many bacteria, it grows to form biofilms, which are resistant to antibiotics. Such biofilms can coat the respiratory and urinary passages of patients with the disease. Two cell-to-cell signal molecules, both homoserine lactones, have been shown to regulate virulence genes through a quorum-sensing mechanism and have been identified in the biofilms. Focusing on the roles and on the control of these autoinducers is a new approach to understanding biofilm formation in cystic fibrosis patients.

But let's go back to bioluminescence. If a few bacteria found themselves in an enclosed cavity in which they were provided with nutrients, would they not colonize it and in gratitude turn-on their not-yet-induced light-emitting system? This is apparently exactly what happens in the case of a small squid, *Euprymna scolopes*, also called the Hawaiian bobtail squid, and its luminous *Vibrio fischeri* guests. This squid (about 3–4 cm long, Figure 5.6), active at dusk and nighttime, can be found in abundance in shallow waters off the Hawaiian coast. It can be bred and raised in the lab, and it has truly become the poster child for a successful mutualistic symbiosis. Because both partners can live independently, their partnership has taught us unique lessons about the initiation and sustenance of their lives together. What the bacteria gain is room and board, and what the squid gains is a lantern, ventrally located, which it uses for counter-illumination to match the light from the moonlit sky and become less visible to potential predators below.

Figure 5.6. The squid *Euprymna scolopes*. The ventral light organ is located in the mantle.

A freshly hatched squid, emerging from twenty days of incubation, gives out no light because it has no symbionts. During embryogenesis, the squid develops a rudimentary organ ready to interact with the bacterial symbionts shortly after hatching. Luminous bacteria from the oceanic environment (they number about one per milliliter) aggregate outside of this organ in association with the mucus it sheds and, after a couple of hours, migrate down ducts to the now ready crypts of the light organ. Then, an amazing cascade of events and morphological changes take place on a precise timetable.

It goes like this. Once in these deep crypts, about four hours after the juvenile squids have hatched, the bacteria quickly proliferate. Then, at a critical density, they begin to emit light. This initiates a program of cell death (apoptosis) within the squid, which transforms the morphology of the cells lining the crypts, resulting in the adult light organ. At the same time, 24 hours after entering the squid, the bacterial symbionts have lost their flagella and become smaller.

The light organ of the adult squid is located in its mantle. It is an elaborate, bi-lobed organ with a number of crypts for the bacteria to settle in. It is equipped with a lens, a reflector, and a black diaphragm for the control of the emitted light intensity, since the bacteria emit light continuously. The tissue harboring the symbionts can actually detect light and measure its intensity using, it appears, the same opsin protein responsible for the squid's vision in the retina of their eyes. This fascinating observation provides a possible explanation for the observation that a squid can adjust its ventral emission to match that seen from the sky, as has been shown in this and several other ventrally emitting luminous animals.

How does the squid select *Vibrio fischeri* as the only species among luminous ones colonizing its light organ? This probably depends partly on cell surface antigens characteristic of individual species, as in many other symbioses. But the squid will eventually reject even *V. fischeri* if it does not produce light; this was shown with a dark luciferase mutant, pointing perhaps to another function of the light-sensitive tissues of the light organ. New data are sure to add to this amazing story of the squid's ability to grow a dense monoculture of a specific luminous bacterial strain.

Another question is, how does the squid control its bacterial population, which in the adult light organ reaches about a trillion? It does so quite simply by ejecting 90 to 95 percent of its bacteria every day at dawn, leaving the remaining bacteria to grow, divide, and refill the vacant space. The ejected bacteria join the population of planktonic "free-living" bacteria in the sea but are only potentially luminous, being so dilute that a critical autoinducer concentration cannot be achieved and more luciferase is not synthesized.

Thus, they do not waste energy needlessly on luciferase synthesis and light emission, but they are available to infect the next generation of juvenile squids. It should be noted that while both the light organ and the bacteria undergo significant changes in accommodating to the symbiotic relationship, the bacteria themselves seem to emerge undamaged from their night job in the squid, ready to serve a new host.

Studies of luminous bacteria provide an interesting story with regard to luminous and nonluminous strains, and how squid manage to only culture the luminous forms. Maintained in the laboratory with infrequent sub-culturing, occasional nonluminous colonies will appear (see Figure 5.1), and the nonluminous forms ultimately overgrow the luminous ones. Conversely, cultures of these "dark" strains occasionally give rise to bright colonies. This can be explained by assuming that as luminous cultures grow, a "dark" variant, spared the energy required for light emission, outgrows the luminous form. This might be viewed as a mechanism to test the waters; a nonluminous form might be better adapted to compete in the open ocean where luminescence has no obvious survival value.

The genes for luminescence have evidently not been lost in the dark forms; the occasional luminous form appearing in cultures of dark strains provides the opportunity for the bright strain to prosper if conditions call for it. Without selection pressure for luminescence, dark variants' cells may be selected for on the basis of their more rapid growth. Such bacteria can be viewed as having two life styles, luminous and dark.

Squid are not by any means the only ocean creatures to take advantage of luminous bacteria. Many species of teleost fishes do it also, with great fantasy and humor in the design of their light organs.* At least eleven different taxa have been found to have organs housing symbionts located in different regions of the body, many as intestinal diverticula, and many using *Vibrio fischeri* bacteria, as the squid does (Figure 5.7).

---

* To be sure, many species of both squid and fish have intrinsic bioluminescence systems, not based on symbioses with luminous bacteria.

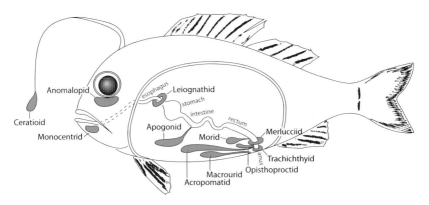

Figure 5.7. An illustration of a generic teleost fish used to show the locations (*in green*) of light organs hosting luminous bacteria in eleven different named taxa.

The tropical fish *Leiognathus nuchalis*, dubbed the pony fish because of the appearance of its mouth when feeding (Figure 5.8), harbors a pure culture of a different species, *Photobacterium leiognathi*, in an appendix-like organ connected to the esophagus (Figure 5.9); with growth, excess bacteria are released into the gut tract and expelled in fecal pellets. The continuous light emission first enters the internally reflecting swim bladder and is then transmitted via fiber optic–like tissues to the ventral side of the fish, where it serves to conceal the silhouette from predators below, just as the little squid does. The intensity of the emitted light is also regulated to match that of the down-welling light.

A specific group of tropical reef fishes, such as the flashlight fish *Photoblepharon steinitzi* (Figure 5.10), have large lanterns under the eyes; these organs are chock-full of bacteria, packed in rows separated by channels guiding the blood vessels that bring nutrients and oxygen (Figure 5.11). The chambers communicate directly with the outside via surface pores; as the bacteria grow, the excess are expelled daily into the ocean. The fish can turn their headlights on and off using lids external to the light organs. Watching the fish swim straight ahead with its headlights on, then off when it makes a sharp turn, taking on a new trajectory, is a spectacle to behold. More on the flashlight fish in Chapter 6.

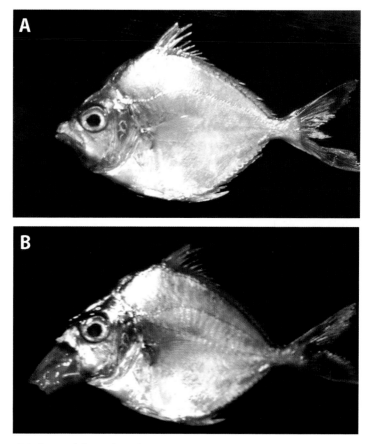

Figure 5.8. A pony fish purchased at an open market in Ambon, Indonesia showing its appearance as when purchased (A), and after having its mouth forcibly opened (B).

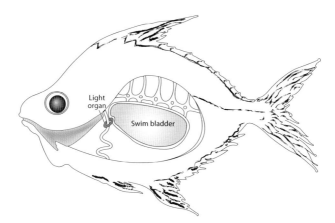

Figure 5.9. Drawing of the pony fish *Leiognathus nuchalis*. From the organ (*green*), light enters the swim bladder, which is internally reflecting. Light is then conducted by fiber optic–like tissues to the underside of the fish.

Figure 5.10. The flashlight fish *Photoblepharon*, with its large suborbital light organ.

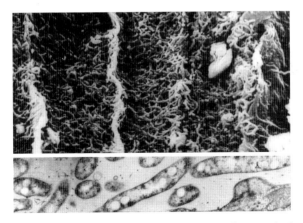

Figure 5.11. View of the *Photoblepharon* light organ (top), showing the luminous bacteria densely packed in compartments. Electron microscope image of sectioned bacteria (bottom).

Although many deep-sea fishes use a coelenterazine/luciferase system to luminesce, the anglerfish (Ceratioidei) have more wit. They pack luminous bacteria in little lures hanging in front of their mouths (Figure 5.12). Apparently only the females were ingenious enough to come up with that idea. In fact, the males of these species

Figure 5.12. The anglerfish *Melanocostus*, with a parasitic male attached (ventrally) and a lure with luminous bacteria on its esca, above.

are often hapless tiny creatures reduced to end their lives fused to the abdomens of their mates—more on this in Chapter 6 on the deep ocean.

Most luminous bacteria species belong, without a doubt, in a chapter on marine bioluminescence. But there are a few exceptions. *Photorhabdus* (*Xenorhabdus*) strains have been found, albeit rarely, in human wounds, and other strains of *Photorhabdus* are symbionts of nematodes that, in turn, infect insects, rendering them bioluminescent . . . on their way to death (see Chapter 6). In addition, strains of *Vibrio albensis* (a freshwater species closely related to *Vibrio cholerae*) have occasionally been isolated from riverine environments.

Before leaving the bacteria to their tricks, let's go back to 1870 and a page of Jules Verne in *Twenty Thousand Leagues under the Sea:*

> About seven in the evening, the *Nautilus*, half-immersed, was sailing in a sea of milk (in the Bay of Bengal). At first sight, the ocean seemed lactified. Was it the effect of the lunar rays? No: for the moon, scarcely two days old, was still lying under the horizon in the rays of the sun. The whole sky, though lit by the sidereal rays, seemed black by contrast with the whiteness of the water.

Such eerie experiences have, in fact, been reported many times over the centuries, mostly from merchant ships sailing in the Indian Ocean; there were over 200 reports of such sightings in the twentieth century alone recorded in ship logs deposited in archives in London; scientists paid little or no attention.

But in 2005, scientists at the Monterey Bay Aquarium and at the Meteorology Division of the Naval Research Laboratory, both in Monterey, California, had the brilliant idea of matching a first-hand account of a milky sea from a ship crossing in the Indian Ocean east of the Somalian coast with satellite images of the area on the same night. The results were stunning. On the night of January 25, 1995, the captain of the British SS *Lima* wrote that "the bioluminescence appeared to cover the entire sea area . . . as though the ship was sailing over a field of snow or gliding over the clouds," and recorded the latitudes and longitudes when they first encountered and left it. Indeed, the satellite images of that same night and two subsequent nights clearly show a luminous area, about the size of Connecticut, exactly where and when the ship traversed it (Figure 5.13).

Do we know the source of such extraordinary seas of bioluminescence? Not yet with certainty. But expert fingers point to the luminous bacterium *Vibrio harveyi*, perhaps in association with blooms of microalgae, allowing a necessary autoinducer to accumulate. If so, the

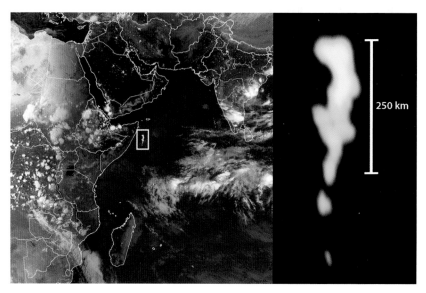

Figure 5.13. *Left.* The region where a milky sea was spotted (*box,* color enhanced) by a merchant ship in the Indian Ocean near the horn of Africa in 1996. The exact longitudes and latitudes when it entered and exited the area were recorded; an archival image from a satellite camera on that night at the same time provided the image shown (enlarged, right) and confirmed the coordinates exactly.

light would be bluish-green; the satellite was, of course, color-blind, and milky glows are probably so weak that they appear white to the rod vision of night sailors. More sea captains with a naturalist bend are definitely needed to broadcast their sightings and attract swarms of microbiologists to the milky waters.

part two

DIVERSITY, FUNCTIONS, AND EVOLUTIONARY ORIGINS OF BIOLUMINESCENCE

The field of bioluminescence is, in many ways, a curio shop of folk art, full of beautiful light works of still mysterious origin. In the course of Part I of this book, where the focus was on the best-understood bioluminescence systems, we allowed ourselves digressions to tell of animals that have co-opted such model systems. The little squid *Euprymna scolopes*, with its ventral pouch full of luminescent bacteria, is one example. This is truly a fascinating story, with exciting new developments around the corner.

How the flashlight fish *Photoblepharon* maintains its culture of luminous bacteria in the pouches under the eyes is yet to be fully understood. In total darkness, the light attracts small phototactic crustaceans and allows the fish to see and capture them, while the fish avoid predation by the blink-and-run technique. And still on the topic of bioluminescent bacteria, can Milky Seas be demonstrated to be of bacterial origin?

The remarkable association between mound termites and beetle larvae in Brazil deserves more than the few lines it gets in this book, as well as an increased effort in the field to preserve these mounds from farmers' encroachment.

In Part II of the book, we examine the diversity, functions, and evolutionary origins of bioluminescence. Chapter 6 brings together, in no particular order, ten striking examples of luminous organisms occupying various but quite specific niches, where the biochemical bases of emission are mostly not yet well established. Many, or perhaps most, can be expected to have originated independently in evolution and thus to rely on different luciferins and luciferases. The Oceans, the subject of Chapter 7, are perhaps where bioluminescence finds its true raison d'être, probably a consequence of evolutionary history (see Chapter 9). After an essay in which we consider in perspective all the various functions of bioluminescence (Chapter 8), most of which we have already encountered somewhere along our journey so far, we conclude in Chapter 9 with an examination of an earlier and speculative but exciting hypothesis—the proposal that bioluminescence originated as a mechanism not to emit light but to remove molecular oxygen from cells early in evolution, when oxygen from newly evolved photosynthetic bacteria was toxic to cells that had evolved in its absence.

Figure 6.1. As one looks out over the Gulf of Aqaba from Sinai, the light from scores of flashlight fish illuminates the water in discrete areas. The photograph is a double exposure: the first was taken at dusk, showing Saudi Arabia on the horizon and the Crusader Island in the foreground; the second is a time-exposure (~40 min) at night, showing streaks in the sky from the movement of stars.

chapter six

SHORT ACCOUNTS OF OTHER LUMINOUS ORGANISMS
Having different and not well-characterized biochemistries

Light for All Reasons: The Several Functions of the Flashlight Fish Light Emission

After the Six-Day Arab–Israeli War in 1967, soldiers on patrol along the east coast of the Sinai Peninsula noticed spots of light close to shore in the Gulf of Aqaba (Figure 6.1). Thinking it might be scuba divers from Saudi Arabia, just across the Gulf, they ordered depth bombs to be dropped, only to see hundreds of small stunned fish come to the surface, with bright light emitted from organs beneath their eyes.

At that time there were no paved roads, buildings, or artificial lights in the Sinai. It may have had fewer inhabitants—most were Bedouin nomads with camels and tents—than in biblical times, when Moses and the Israelites wandered in the area for 40 years after escaping bondage in Egypt.

This was the first description of the fish, *Photoblepharon steinitzi*, now called flashlight fish, in the region of the Red Sea. The related species *Photoblepharon palpebratus* had long been known and well described from the waters in Indonesia, but *P. steinitzi* was not listed in the voluminous *Fishes of the Red Sea*, possibly because these fish never venture out during the day, only on completely dark moonless nights; at other times, they may stay in the dark somewhere beneath the coral reefs. Other related species have since been discovered in the Caribbean.

In subsequent years, scientists found *P. steinitzi* to be very abundant in the Gulf of Aqaba region, where they could readily be seen and photographed by snorkeling (Figure 6.2), and even captured by hand for transfer to aquaria in the laboratory. They were described

Figure 6.2. A group of flashlight fish shown over a reef in the Gulf of Aqaba, south of Elat. The number of fish in a group is sufficient to create a glow in the water as seen from shore (Figure 6.1).

in *National Geographic*, along with a striking photograph of their light as seen in the sea; several public aquaria in the United States have exhibited them.

Located just below the even larger eyes, the large paired light organs culture luminous bacteria, which emit light continuously; the emission is controlled with lids that move upward to cover the light organ. Although the organs are located like headlights (Figure 6.2), the fish were called flashlight fish because, like flashlights, they have several evident uses. Their light attracts phototactic zooplankton such as small crustaceans and allows the fish to see and capture them. The fish avoid predation by the blink-and-run technique; they turn the light off for an instant, change swimming direction, and scoot off in the dark. They also use the light for communication between sexes, as inferred by behavioral patterns.

Mating in a Circle of Light: The Bermuda Fireworm (*Odontosyllis enopla*) and Other Bioluminescent Marine Annelids

In the case of several bioluminescent organisms, we know something about how they emit light, but we do not really know why (bioluminescent mushrooms come to mind); in many other cases, we may not know exactly how they do it, but we certainly know why. The *Bermuda fireworm* and its allies (Figure 6.3) most definitely belong to the latter group: they do it to attract mates by following the most astonishing of scenarios.

Consider this: during the summer months, two or three nights after the full moon and on several consecutive nights, about 57 minutes (between 51 and 63 minutes, to be precise!) after sunset, the females (about 2 cm long) leave the bottoms of the shallow coves and inlets where they live and swim to the surface; there, they engage in a frenzied and spectacular bacchanal, each female swimming in a luminous tight circle (Figure 6.3A). The males respond by darting from below to the center of a circle, joining in the carousel, and emitting bursts of light. Males and females then circle together, the males shed sperm and the females release eggs along with a brilliantly bluish-green luminescent slime. If no male shows up, the female momentarily stops her circular dance, then resumes it. Other species around the world produce variations on this theme, and most are equally precise.

Fertilization occurs in the seawater, and success is apparently excellent. This evidently requires that the males be able to see the females' light. Although males are only half the size of the females, their four eyes are nearly double the size of the females' (both sexes have four eyes). Moreover, their visual response is spectrally attuned to the bioluminescence peak (~515 nm). The chemistry of

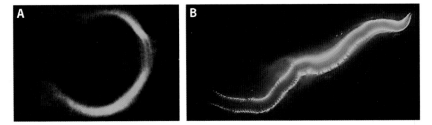

Figure 6.3. The Bermuda fireworm, *Odontosyllis enopla*. A worm, female, during a mating dance at sea (*A*), and another worm, also female, in seawater in a dish in a photo taken at the site (*B*).

the light-emitting system, which involves a luciferin and a luciferase, has not yet been worked out in spite of serious attempts.

That the Bermuda fireworms are a tourist attraction is not a surprise. Similar polychaetes can apparently be observed throughout the tropics; the story goes that Christopher Columbus witnessed a display on his first voyage.

Another species of bioluminescent polychaete worm, *Eusyllis blomstrandi*, apparently does not engage in flashy mating rituals like its close relative and look-alike, the Bermuda fireworm; it emits repeated flashes, evidently as a way to deter predators. Easily collected around the island of Helgoland in the North Sea, it lives in home-spun silk tubes in the thalli of red algae and can be kept alive and well for months in the lab, where high-speed video recordings of its emission can be obtained under the microscope, on the msec time scale. Such movies show how the luminescence propagates along the length of the worm when touched. Under strong stimulation, the entire trunk of the worm luminesces as a series of flashes originating from symmetrical patches on each segment of the worm. Each of these patches is apparently composed of smaller light sources, and the flashes propagate along the worm at a speed greater than 1 msec per segment.

When the worm is touched hard, it will apparently commit spontaneous "autotomy"; that is, it will break itself in two. The anterior part, which can regenerate a tail and survive, emits a few flashes before quickly leaving the scene of the accident with all lights turned off. The helpless posterior part, which cannot regenerate, stays behind at the scene, doomed to starvation. There it glows very brightly for nearly a minute, as if acting as a decoy to give a chance to the head fragment to escape the predator and survive, the familiar lizards' tactic. The polynöid polychaetes (scale worms) engage in a similar tactic, except that they shed glowing scales as a diversion.

If the purposes of the bioluminescent displays in these two species seem obvious, one is at a loss to figure out why the polychaete marine annelid *Chaetopterus variopedatus* would bother to emit light.

Indeed, this worm, abundant in many places of the world, constructs and lives its entire life in a U-shaped parchment tube, buried in the mud near shore, with only the two well-concealed ends of the tube open to the ocean. But it has been noted that many small commensals invade the tubes, and that the light emission may serve to divert these intruders. The worm's body is adapted to tube life, with appendages in its mid-section that look and act like paddles and serve to move water carrying oxygen and nutrients through the tube. The worm can be gently removed from its tube and placed in a dish; when disturbed, the posterior third of the worm emits a flash of blue light and secretes a luminous slime. The blue emission peaks at 457 nm.

Very little is known of the chemistry of this emission, except that it requires oxygen, as all bioluminescences do. Some 40 years ago a large protein, probably polymeric (i.e., made of repeat units), was isolated and shown to emit light in the presence of ferrous ions and hydrogen peroxide, in proportion to the amount of protein. Considering that *Chaetopterus* has long been a guinea pig in cell division research and embryology, it is curious that its bioluminescence has attracted so little interest in recent years.

Ctenophores, Beautiful by Both Day and Night: Why Are Most Species Bioluminescent?

While swimming in shallow coastal waters in most parts of the world, one may often feel, but not readily see, a jelly-like object against one's body. These are jelly-like animals, so-called sea walnuts or comb jellies; on some occasions, they may occur in massive numbers, a virtual flotilla of jellies bobbing in the waves. All are ctenophores, members of the phylum Ctenophora.

When stimulated at night, ctenophores emit brilliant flashes—said to be the brightest of all luminous emissions. These originate from photogenic cells located along eight bands, called comb plates or paddle plates, arranged circumferentially like the seams in a football (Figure 6.4). By day the plates exhibit a striking iridescence, sometimes mistaken for bioluminescence. The plates are rows of fused cilia, used for locomotion, which is slow at best, but unrelenting.

Near-shore species of the *Beröe* and *Mnemiopsis* genera have been more studied than open- or deep-ocean species, but ctenophores do

Figure 6.4. A ctenophore by daylight; the comb paddles are girdles running longitudinally, some of which exhibit the iridescence; flashing occurs from photocytes located along the paddles. The simple opening at the bottom is the mouth.

occur worldwide in all latitudes and depths of the ocean. Unlike most animal phyla, which number thousands or millions of members, there are only about 150 known Ctenophora species, all marine; also unlike other taxa, most—maybe more than 90 percent—are bioluminescent.

If most ctenophores are indeed luminous (only one exception has been reported so far), what might be the selective advantage for ctenophore bioluminescence, such that light emission is retained in so many members? The function of light flashing in ctenophores is widely believed to have a role in deterring predation, as in many other species. But can flashing be that advantageous? Why not remain dark all night and rely on near-transparency during daylight?

For 50 years ctenophores had stood as an unexplained exception to the rule that all bioluminescent organisms require oxygen; ctenophores in seawater from which all oxygen was removed emitted undiminished flashes. The discovery of the aequorin system in jellyfish explained this: the enzyme-coelenterazine-peroxy intermediate is stored in quantity in photogenic cells. The final steps, triggered by calcium, require no oxygen. Ctenophores, related to coelenterates, were shown to have a similar biochemistry.

An unusual feature of the ctenophore system is its inactivation by exposure to light and the absence of circadian control. Bioluminescence is emitted only when the animals are kept in the dark; exposure to light during the night phase suppresses luminescence, and animals in daylight will become luminous again within about 30 minutes if placed in the dark at any time of day. The light exposure causes the luciferase system to be photoinactivated, the molecular target very likely being the peroxy intermediate.

Many ctenophores are voracious predators, indiscriminately gulping in organisms ranging from small animal larvae, phyto- and zooplankton, to small crustaceans; they may ingest many times their own weight in a day. This habit wreaked havoc with the then-thriving fisheries in the Black Sea when the common American species, *Mnemiopsis leidyi*, was introduced from ballast water in the early 1980s.

Its diet included developing fish eggs, such that the fish population and fisheries completely collapsed. The later introduction of the invasive ctenophore *Beröe ovata*, a specific predator of lobate ctenophores, seems to have brought *Mnemiopsis* under control. Meanwhile, the *Mnemiopsis* have moved on to the Caspian Sea, where a similar scenario has unfolded. Ctenophores are beautiful, but they have proved themselves to be one of the most damaging of any invasive species to an ecosystem.

A Clam Lodging in a Rock: *Pholas dactylus*

Known since antiquity, the mollusk *Pholas dactylus* is a clam that is abundant along the shores of England and France. It uses its sharpened valves to bore into soft rocks and lodge there (Figure 6.5); the blue-green bioluminescence (λmax ~ 505 nm) is associated with symmetrical patches along the body of the clam and with a luminous mucous excreted through the siphon, perhaps frightening predators. In truth, the actual function remains enigmatic.

Pholas has a very special place in the story of bioluminescence, because it was the first organism from which a cell-free luminous extract was obtained by Dubois in 1885. Later he found that both *Pholas* luciferase and luciferin were proteins, differing only slightly in their heat stability; after 3 min at 65°F luciferin retains 40 percent of activity, while the luciferase is fully inactivated. He also reported that the "luciferin protein" could be oxidized with light emission not only in the presence of luciferase but also by a variety of oxidizing reagents, such as permanganate, hydrogen peroxide, and hypochlorite.

These intriguing results were confirmed and much expanded on in the 1970s. *Pholas* luciferase is now known to be a glycoprotein of 310 kDa containing two copper atoms and consisting of two subunits of 150 kDa each. The luciferin, now named pholasin, is also a glycoprotein (34 kDa), with a small tightly bound prosthetic group (the true luciferin?) necessary for bioluminescence emission. Luciferase

Figure 6.5. *Pholas dactylus* in a rock cavity bored by the clam itself; removing the clam requires breaking the rock. The clam is a delicacy on the table; when eaten in a dimly lit room, the bright luminescence can be seen through the cheeks.

has two independent binding sites for pholasin, each probably associated with one of the subunits. Luciferase and pholasin form a 1:2 complex and can be separated by gel filtration.

What makes the *Pholas* system special is that both luciferase and pholasin show nonspecific activity. Luciferase acts as an oxidase with pholasin, catalyzing its oxidation and light emission in the presence of oxygen. As Dubois also described, luciferase can also act as a peroxidase with substrates other than pholasin, such as ascorbate and H_2O_2, albeit without light emission. Likewise, pholasin alone can emit light upon reaction with a variety of reagents that generate reactive oxygen species (ROS).

Pholasin, extracted from the clam, is actually available commercially as a luminescent reagent for the detection of free radicals and reactive oxygen species. The apoprotein has been cloned and expressed, but the chromophore, essential for emission, has so far eluded definitive identification. Recent evidence points to dehydrocoelenterazine, which is also the bioluminescence chromophore of another mollusk, the squid *Symplectoteuthis oualaniensis,* as an attractive possibility.

The Only Known Luminous Freshwater Animal: The Limpet *Latia neritoides* of New Zealand (A Salty Question!)

Why are the world's ponds, rivers, and lakes—some large, deep, and dark, such as Lake Baikal in Siberia—not home to a rich fauna of bioluminescent organisms, like the oceans? Evolutionary biologists believe that life originated and went through its early major evolutionary steps in the sea. But once established on land and freshwater, animals and plants experienced an expansive development into diverse habitats. One would think that estuaries would have provided unlimited opportunities for mutations from salt-dependency to have arisen. But no such evolution seems to have occurred in bioluminescent systems. For example, no bioluminescent representatives of any phylum except *Latia* have been found in lakes around the world, not even in Lake Baikal.

It cannot be simply the salt dependency. The luminous bacteria from human wounds and those that parasitize caterpillars (*Photorhabdus*) grow well on low-salt medium but poorly on seawater; luminous mushrooms, worms, and beetles exhibit no special salt requirements.

This leaves us to marvel at the only known freshwater exception, a very small snail named *Latia neritoides,* less than 1 cm in diameter (Figure 6.6). *Latia* is found exclusively in some of the shallow rivers and fast-running streams of the North Island of New Zealand. It clings to the underside of rocks, making it hard to notice. But if touched, it releases a green-glowing sticky mucous ($\lambda_{max} = 536$ nm), which could well startle potential predators. If a predator does come into contact with the sticky slime, it may take it hours to get rid of it, making it now more evident to its own predators. The luminescence in this case may thus serve as a warning of something to be avoided, and thus protect *Latia.*

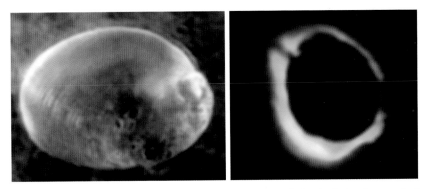

Figure 6.6. *Latia neritoides*, shell size ca. 7 mm (left). At night by the light of its exuded mucous (right).

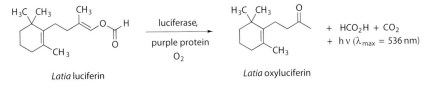

Latia luciferin   *Latia* oxyluciferin

Figure 6.7. Chemical structure of *Latia* luciferin and its postulated reaction scheme.

The bioluminescence chemistry of *Latia* has been partly worked out (Figure 6.7). It involves a large glycoprotein luciferase (~180 kDa), a luciferin of identified and unique chemical structure, and oxygen. A second protein present in the bioluminescent mucous, called the purple protein (~38 kDa), has been implicated in the reaction, but it is not needed for emission and its fluorescence is red, not green; it may nevertheless be involved in the reaction in some way.

The mechanism of the light emission remains a mystery. The identified products of the enzymatic oxidation of *Latia* luciferin are CO_2, formic acid, and oxyluciferin, but the stochiometry of the reaction has not been established. Oxyluciferin itself could not be the emitter, since it absorbs only in the UV and is not fluorescent, but it might be when protein bound. Synthetic analogs of luciferin can be substituted for the natural luciferin without modifying the bioluminescence spectrum. Fingers point to the possibility of an unidentified chromophore tightly bound to the luciferase, perhaps a flavin, whose fluorescence spectrum is a good match to *Latia*'s bioluminescence.

But the fluorescence spectrum of this luciferin-luciferase system reveals only the weakest evidence of flavin, or of any other chromophore. A process of energy transfer from excited oxyluciferin to a fluorophore tightly bound to luciferase, but present only at very low concentration, would have to be exceptionally efficient. The only freshwater bioluminescent organism deserves more attention.

A Luciferase Reaction That Requires Hydrogen Peroxide, H_2O_2: *Diplocardia longa*, a Large Earthworm from the Southern United States

Dozens of species of bioluminescent earthworms have been identified; they are worldwide in distribution, ranging from a millimeters-size species in Siberia, to very large ones in Australia and the southern United States (as long as 2 feet). *Diplocardia longa*, a species that is abundant in regions of the U.S. state of Georgia is about 10 inches in length. Like *Chaetopterus*, the function of the light emission is a matter for speculation; experiments are needed.

The luminescence occurs in an exudate of coelomic fluid containing large cells, identified as being responsible for light emission. The fluid is ejected from the mouth, anus, and dorsal pores of the animal, but only following vigorous stimulation. The body of the worm itself does not emit light and appears in silhouette against a background of luminous slime (Figure 6.8). Indeed, the cells themselves do not emit unless and until broken.

The system differs biochemically from all others described in this book because *in vitro* the light-emitting reaction apparently does not use molecular oxygen. Instead, it requires hydrogen peroxide, together with luciferase (a copper-containing protein, molecular mass about 300 kD), and luciferin, an aldehyde. The reaction proceeds unabated in an atmosphere with all oxygen removed, and light emission ceases if H_2O_2 is removed (e.g., by an enzyme that destroys it).

But oxygen is required for emission in the exudate of broken cells; if a worm is placed in an atmosphere without oxygen and stimulated vigorously, the coelomic fluid is discharged and the cells broken, but

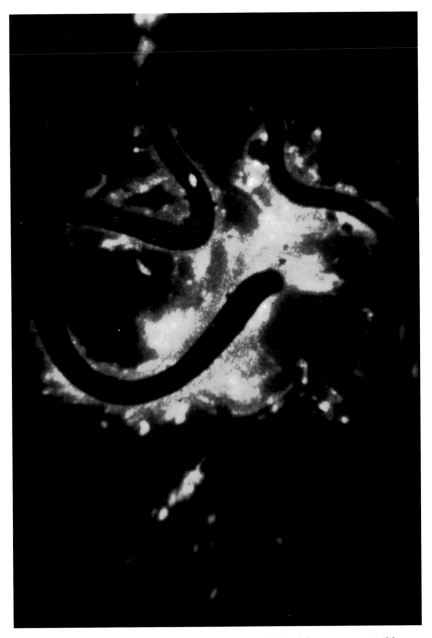

Figure 6.8. Luminescent slime from an earthworm, with the slithering worm visible as a silhouette.

no light emission occurs. If oxygen is then admitted, a brilliant bluish luminescence is observed. The assumption is that the cell lysate releases and activates an oxidase enzyme system that requires molecular oxygen and produces hydrogen peroxide. Studies of earthworm biochemistry thus concluded that the luciferase is a peroxidase that oxidizes the aldehyde to the acid, but researchers are silent on the identity of the emitter; there are no indications of fluorescent molecules in the purified luciferase and the synthetic luciferin.

While this is the only system shown not to use molecular oxygen directly in the luciferase reaction, there may be others, such as *Chaetopterus* and scale worms, where a different active oxygen form may be used.

Bioluminescent Bacteria on Dry Land: *Photorhabdus luminescens* in Human Wounds and in Insects via Symbiosis with a Roundworm

Toward the end of Chapter 5, we mentioned the only known terrestrial species of luminescent bacteria, in the genus *Photorhabdus* (formerly *Xenorhabdus*); on a few occasions, strains of this group have been isolated from human wounds. Other strains, much more common, parasitize and kill insects with the help of nematodes in the genus *Heterorhabditis*. The study of nematodes is an active research area, as a quarter of the world population is infected with nematodes.

The "phosphorescence" of wounds had long been described and regarded as beneficial; wounds infected with luminous bacteria were said to heal faster—a far cry from the truth, it now appears. Around 1990 a strain of luminescent bacteria was isolated from a human wound and characterized. It was exciting to find that the optimal temperature was 33°C for its growth and 40°C for bioluminescence, clearly pointing to an evolutionary association with warm-blooded (mammalian) hosts. In other respects, the properties of this *P. luminescens* luciferase do not differ much from those of its luminescent marine cousins. It has two subunits, α and β, and the reaction requires a long-chain aliphatic aldehyde; an autoinducer is needed for the synthesis of luciferase and the development of luminescence, so bacterial populations must reach a certain density before luciferase and related proteins are synthesized.

An Australian report in 2000 on four cases of human infections makes *P. luminescens* seem unattractive indeed. In two of these patients, the lesions appeared to follow spider bites, as in one of the earlier American cases (the role of the spiders has not been investigated). Though the illnesses started in all cases with painful localized lesions, they became systemic and took weeks to finally be cured with appropriate antibiotics. The moral of the story is clear. In case of lesions caused by unidentified bacteria, a dark room is the place to go; if the wound emits light, the culprits are *P. luminescens*.

Many related strains of *Photorhabdus luminescens* have been isolated from associations with nematodes or from a wide range of nematode-infected insects. But they have no altruistic intentions; on the contrary, they bring the insects to a brilliant death while enhancing their own propagation. To this murderous end, the pathogenic bacterium, unable to penetrate the insect by itself, exhibits a sophisticated mutualistic association with its nematode host in a two-stage lifecycle (Figure 6.9).

From birth, juvenile soil nematodes (Figure 6.9) carry the bacteria in a small oesophagal pouch, where they neither multiply nor luminesce, but wait. Finding a caterpillar (the larval form of a butterfly, for example), a nematode enters its gut tract, penetrates the gut wall, and gets into its body cavity, where it releases the bacteria, which grow rapidly on the rich nutrients in the hemolymph. The bacteria secrete proteases in the hemocoel that degrade the insect proteins, as well as high-molecular-mass toxins that kill the insect. The nematodes, feeding on the rapidly multiplying and now-luminescing bacteria, as well as on the dying or dead insect, mature to a hermaphroditic stage and produce eggs, which develop into male and female infective juveniles. Finally, when the food supply is exhausted, these fresh infective juvenile nematodes (as many as a thousand, all carrying symbiotic bacteria) emerge from the cadaver into the soil (sometimes a distance away, if moved by a scavenger that consumes the caterpillar cadaver). They locate a new caterpillar, and

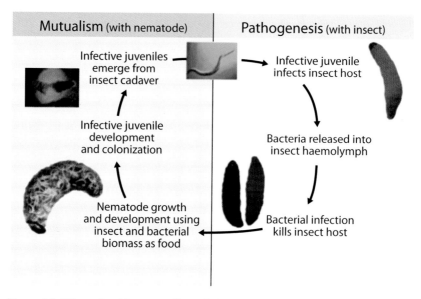

Figure 6.9. Life cycle of luminous *Photorhabdus* bacteria, which colonize the gut of infective juvenile nematodes that, in turn, infect a caterpillar (insect larva) and kill it.

a new life cycle starts again. The bacteria need the nematodes to get into the insects, and without the bacteria—without the *specific* bacteria—the nematodes could not kill the insects or complete their life cycles. Nonluminous bacteria also infect nematodes/caterpillars, and both luminous and nonluminous infective bacteria produce a red pigment.

What is the point of the luminescence and of all the enzymes associated with it in this complicated symbiosis? Who would want to feed on a luminous dead caterpillar? A nocturnal animal? No one has a good answer yet. The complex tripartite association nematode—bacteria—insect is "mutualistic" for the first two parties, but only pathogenic for the insect caterpillar.

Dipteran Glowworms: Fly Larvae in Caves and Other Damp Places

Less known than bioluminescent beetles, whose larvae are also called glowworms, the dipteran glowworms of the genera *Arachno-*

campa (Australia and New Zealand) and *Orfelia* (Appalachian Mountains, United States) are intriguing insects. The adults of these two genera look much like mosquitoes (Figure 6.10A), have short life spans, emit little if any light, and are of no great interest for this account. But their long-lived larvae are extraordinary (Figure 6.10B). At first glance, they look very similar, about 10 to 20 mm long and worm-like. Both species inhabit dark, humid places, often on the slopes of riverbanks, although *Arachnocampa* favors caves. Both build webs, both emit blue light, and both are carnivorous, trapping and eating small insects attracted to their lanterns. They are even cannibalistic, if deprived of more appealing food. In the isolation of caves or in laboratory constant conditions, the emission from both of these diptera exhibit circadian rhythmicity.

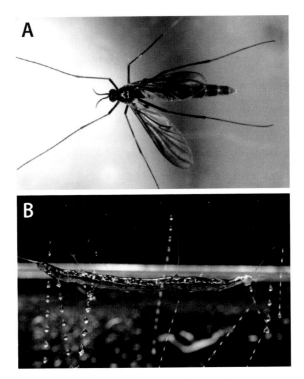

Figure 6.10. *A.* An adult insect *Arachnocampa*. *B.* A larval *Arachnocampa* in its suspended tube and "fishing lines."

But the similarities end there. The larvae of *Arachnocampa*, which emit continuously at night, build and move around in mucous tubes, which they suspend by threads to the walls or ceiling of their dim habitat. From this tube, they produce and hang "fishing lines" carrying droplets of a sticky mucous on which their prey get caught (Figure 6.10B). They then haul up the line and eat the prey in the safety of their home tube. The Waitamo and Te Anau caves of New Zealand are extraordinary sights and money-making tourist attractions (Figure 6.11).

The American *Orfelia*, in contrast, is infinitely more discreet and, in fact, very hard to find. It anchors a strong central stretch of silk by several finer lines to the walls of a crevice where it chooses to live. Poised on this highway, it glides forward and backward with remarkable speed to catch prey trapped in the sticky web, or to avoid a predator (Figure 6.12).

The light organs of these two diptera are totally different. In *Arachnocampa*, the organ derives from the Malpighian tubules and is located at the tail. In contrast, *Orfelia* has two lanterns of very unusual structure, one anterior and one posterior. Both are characterized by two parallel rows of "black bodies" consisting of large binucleated cells filled with dark granules, both of still-unknown function (Figure 6.12). Although both *Arachnocampa* and *Orfelia* emit blue light, their spectra are not identical; *Arachnocampa*'s emission peaks at 484 nm and *Orfelia*'s at 460 nm, the shortest wavelength of any known terrestrial bioluminescence.

The reaction chemistries are also clearly different. *Arachnocampa*'s is an ATP-dependent system; *Orfelia*'s is not. Indications are that the luciferase of *Arachnocampa* may be related to beetle luciferase, which would be of great interest, but it does not use beetle luciferin. In addition to a luciferase, the bioluminescence of *Orfelia* may involve a large luciferin-binding protein. The luciferases of *Arachnocampa* and of *Orfelia* have yet to be characterized, and the same is true of their luciferins. In the case of *Orfelia*, it is difficult to collect enough animals without destabilizing a small population.

Figure 6.11. Tourists in a cave in New Zealand viewing the many luminous fly larvae on ceiling.

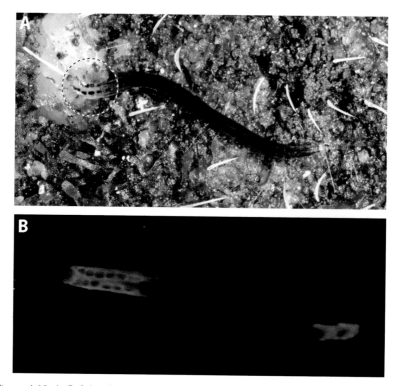

Figure 6.12. A. *Orfelia* glowworm (*diagonally in the picture*) clearly showing the double row of "black bodies" in its head segment (*circled*). *Orfelia* glides on its track to reach prey caught in its sticky web. B. *Orfelia* glowworm at night showing its glowing head and tail segments, with the "black bodies" silhouetted.

A Cyanide-Producing Glowworm on the Forest Floor: The Millipede *Motyxia* (Formerly *Luminodesmus*) *sequoiae*

In 1949, biology students at the then-new campus of the University of California at Santa Barbara took a field trip to the beautiful Sequoia National Forest in the Sierra Nevada range. Bedded down on the soft lawn of needles, one of the students awoke at night and saw all around him spots of light on the forest floor. He awakened others, and they discovered that the light was coming from millipedes, about 4 cm long, glowing continuously over their entire bodies. They returned to campus with some worms in jars and showed them to a faculty member, who sent them to an expert for identification. The animal turned out to be a new species and was assigned the name

Luminodesmus sequoiae (Figure 6.13), now changed to *Motyxia sequoiae*. These blind millipedes, which are also called glowworms, spend day times hidden under leaves and emerge to forage for food at night.

What could be the function of this luminescence? Measurements in the laboratory showed that if the millipedes are undisturbed, the light is fairly constant for hours. But if disturbed, the intensity promptly increases to twice its steady value, slowly returning to the steady level within a few minutes. As in the case of nonluminous

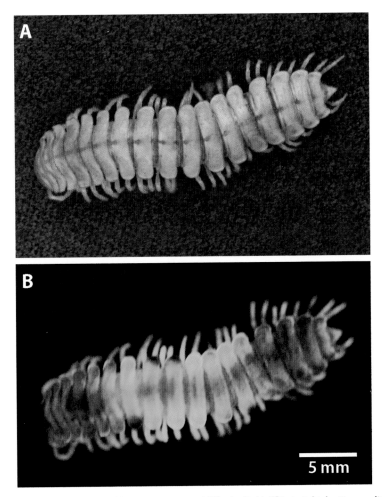

Figure 6.13. The millipede *Motyxia sequoiae* (A) in daylight; (B) at night, by its own light.

members of the Polydesmidae family, disturbance also results in the production of large amounts of hydrogen cyanide, whose strong odor is immediately evident.

Could the bioluminescence be an aposematic signal to warn predators of the cyanide (see Chapter 8)? A recent study, which used hundreds of live millipedes, some with their light sources covered with opaque paint, as well as hundreds of clay models, some mimicking bioluminescence with chemiluminescent paint, showed beyond doubt that rodents avoided the light-emitting millipedes and clay models. Nonluminescent millipedes were attacked twice as often as luminescent millipedes. If a first bite did not kill the attacker, it may remember the light emission and avoid *Motyxia* thereafter. The animals are readily visible and apparently unprotected. Research has given some knowledge as to how these millipedes tolerate their own cyanide, which is a powerful poison for most animals.

Regarding the biochemistry of the bioluminescence, extracts of the millipede have been found to emit light for some minutes, with an intensity that is increased by addition of ATP, but there is no other evidence that the reaction is similar to the firefly system.

Mushroom Bioluminescence: The Whole Mushroom or Only a Part of It May Emit Light

A picture is worth a thousand words, as evidenced by Figure 6.14. Many species of mushrooms all over the world emit dim green light, continuously, day and night. The reader who wonders why is not the first to ask the question; Aristotle did so three centuries B.C.E., and no answer has been definitely accepted since.

One possibility is that insects attracted by the luminescence may help disperse the spores, enhancing propagation. An experiment testing this hypothesis supported it, and most specialists believe that it is the most likely function of the light emission. However, since only the mycelium, which is mostly underground, emits light in the *Armarillia* lineage, not the visible "fruiting body" (the mushroom), this hypothesis, while not negating it, would not seem to apply to

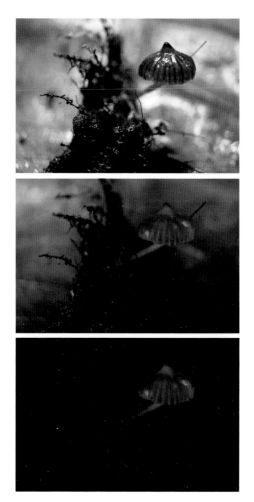

Figure 6.14. The bioluminescent mushroom, *Mycena fera*, which emits light continuously: in full daylight (top); in reduced light illumination (center); in darkness (bottom). This specimen was found in March 2006 on a decaying fern in the Ribeira State Park, Iporanga, SP, Brazil.

that case. Another possibility is that the light serves as warning of distastefulness, thus deterring predators, another aposematic signal (see Chapter 8). Though distasteful, there is evidence that poisonous (to us!) mushrooms do not harm the arthropods and mollusks that eat them.

Be that as it may, what is the biochemistry of the bioluminescence? Since the more than eighty known bioluminescent fungi all belong to the basidiomycetes phylum, and all emit in the same narrow spectral range of 520–530 nm, it is extremely likely that they share the same or similar biochemical pathways. In DuBois-type tests, making use of hot and cold extracts (luciferase and luciferin) from several different species, cross-reactions were positive in both directions.

On this basis, an enzymatic pathway was initially proposed some 50 years ago and has recently received convincing validation. It involves not one but two different enzymes, one soluble in the cold buffer and the other particulate. The role of the first is to reduce the luciferin (L in the scheme below, using NADPH, reduced pyridine nucleotide) present in the hot extract, while the second catalyzes the oxidation of the reduced luciferin by oxygen, which is a step common to all bioluminescence reactions.

$$L + NADPH + H^+ \quad \xrightarrow{\text{Soluble enzyme}} \quad LH_2 + NAD(P)+$$

$$LH_2 + O_2 \quad \xrightarrow[\text{(luciferin)}]{\text{particulate enzyme}} \quad LO + H_2O + h\nu$$

Fungal luciferin has not yet been characterized chemically, but this can be expected in the near future. Forty years ago, in the course of developing a purification protocol for this luciferin (which indeed produced an active crystalline substance), it was reported that light could be elicited in two ways, either via the two-enzyme pathway described above or by addition of hydrogen peroxide and sodium hydroxide. While the enzymatic pathway can be assumed to be responsible for the light emission by mushrooms, the possible role of the chemiluminescence pathway in fungal emission requires further clarification.

The descriptions of these several bioluminescent systems document and underscore repeated themes of the book–namely, that there are many different chemical mechanisms of bioluminescence (e.g., there are many different luciferins and luciferases), that all reactions require molecular oxygen or some form of active oxygen, and that there are many different ways in which organisms use light emission to the common end of providing selective advantage. These functions will be discussed in Chapter 8, and the way in which these different systems may have originated and evolved are considered in Chapter 9.

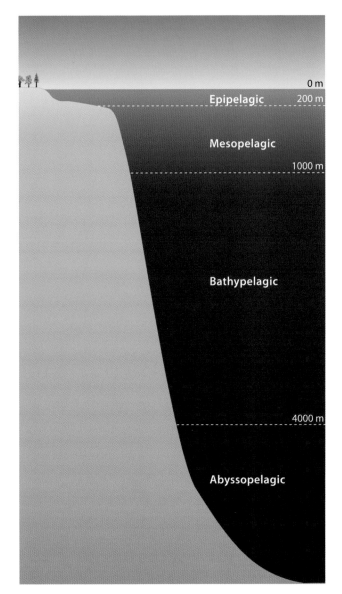

0 m
Epipelagic   200 m

Mesopelagic

1000 m

Bathypelagic

4000 m

Abyssopelagic

Figure 7.1. The vertical zones of the oceans.

chapter seven

BIOLUMINESCENCE IN THE OCEANS
Anglerfish, Dragonfish, and a Lake Baikal Parenthesis

Far away from the beaches, and far down from waves and boats, the deep oceans are forbidding places, places of darkness at noon and numbing cold in all seasons. They are also the vastest spaces on earth; 70 percent of the earth's surface is covered by oceans, 60 percent by deep oceans, a trillion cubic kilometers inhabited by a sparse fauna, a world under tremendous pressure and so dark that no green plant can inhabit it.

It is customary to divide the oceans in vertical zones (Figure 7.1). The upper 200 m of the oceans, the so-called *epipelagic zone*, is full of sunlight and thus of plant and animal life, living in a web of inter-connections and dependencies. At the bottom of the food chain are the phytoplankton, the microscopic photosynthesizers (the dinofla-gellates, diatoms, cyanobacteria), animals (the zooplankton, such as tiny crustaceans) and bacteria. The epipelagic zone is home to many small and large and fast fishes (such as tunas and sharks, flying fishes and their relatives), while many of the luminous species encountered earlier in this book—the dinoflagellates, *Vargula*, *Aequorea*, and some of the fishes that harbor bioluminescent bacteria, such as *Photoblepha-ron*, the pine cone or pineapple fish, as well as the little squid, *Eu-prymna scolopes*, described in Part I—live in the shallow near-shore environment.

The intensity and spectrum of light from the sun changes with depth. In a clear ocean, red light is gradually absorbed by highly excited vibrational transitions (nuclear motions in the H_2O mole-cule), while violet light is scattered by dissolved and suspended ma-terial in the water column. About 200 m down from the surface, the light is blue, peaking around 450 nm, and the intensity is less than a hundredth of what it is at the surface. From the top to the bottom of the next vertical zone (between 200 and 1000 m from the ocean surface), the so-called *mesopelagic zone*, the intensity of ambient light is reduced by more than a trillion!

This weakly blue-lit *mesopelagic* zone is the habitat of abundant and diverse animals, among them many crustaceans, cephalopods, and fishes with large eyes. Many have silvery sides, light undersides, and dark backs, and there are many animals that hide from predators by being transparent or so thin as to be almost two-dimensional. Others hide by using tactics of counter-illumination biolumines-cence, as do the hatchet fishes, with photophores along the ventral region of their compressed body. These little fishes, 2 to 12 cm long, typically live 600 meters below the surface, but their range goes down to 1500 m. Bristlemouths, the most abundant fish in the seas, have two rows of photophores on their underside, while lantern fishes have photophores on the head as well as ventro-lateral rows of them on the body. All these fish engage in a nightly vertical migration, hours long, feeding in the epipelagic zone and even reaching the sur-face, then turning around before daybreak and returning to the depth. Millions of tons of animals engage in such nightly vertical trips, by far the largest migrations of any animals on earth. Now and then a hungry whale feeds on them during their excursion.

Below 1000 m, darkness is essentially total, the temperature never above 5°C (aside from the proximity of hydrothermal vents; see be-low), and food is scarce. This *bathypelagic zone*, which extends to some 4000 m below the surface, and especially the *abyssopelagic zone*, from 4000 m down to the ocean floor, is home to but a few animals, such as squids—giant squids among them—some crustaceans, and some slow-moving fishes, among them the fanciful anglerfishes (see later). Indeed, it has been estimated that over 80 percent of all or-ganisms living in the bathypelagic and abyssopelagic zones emit light, and all fishes from those zones apparently have eyes (even a fish found at 8350 m, *Abyssobrotula galatheae*).

In the *benthic zone*, which refers to the region near the sea floor, irrespective of depth, life picks up again, supported by nutrients and detritus that rain down from above and accumulate. Of the many species of mollusks, squids, fishes, crabs, jellyfishes, sea cucumbers, only a few are luminous. Note that the localization of species into zones is somewhat misleading. Indeed, during the course of their lives, from egg to death, some fishes migrate thousands of vertical meters, from bottom to surface and back, while others, as mentioned above, do not think twice of traveling hundreds of meters up and down simply to catch a meal. The rich biological communities that thrive near hydrothermal vents and black smokers and derive their energy from the oxidation of hydrogen sulfide will not be discussed here, because no example of bioluminescence has been reported from that habitat.

The eye structures of deep-sea fishes, cephalopods, and crusta-ceans reveal diverse and remarkable adaptations to their dark envi-ronment. The stronger-swimming fishes have large eyes designed to detect distant bioluminescent prey, a definite asset where the pick-ings are sparse. Fishes that move slowly have less well-developed vi-sion, because detecting potential prey is of interest only if it can be reached in time. Some fishes near the seafloor, such as the rat-tails, have ventral bacterial bioluminescence organs to help them locate a meal below.

The importance of bioluminescence in the deep ocean might find strong support in the distribution of the peaks of visual sensitivity of 175 species of deep-sea fishes, which fall in the wavelength range of 468–494 nm (Figure 7.2). This range compares remarkably well with the wavelength range of fish bioluminescence, but it is a poor match to the blue-shifted spectral range of the down-welling day-light reaching these fishes (see horizontal bars in the figure). At first glance, the data appear very convincing. There is also a hint that the deeper the habitat, the bluer the λ_{max} of a fish's vision; this relation-ship was, however, not confirmed in a different study of fifty-four lantern fish.

Is this the result of an evolutionary adaptation of species' eyes to the growing importance of bioluminescence with depth? Or could it possibly be a red herring and, in truth, the other way around? Did bioluminescence find its niche in the color-blind world of deep-sea fishes, and its emission spectra adapt to match the retinal pig-ments of the fishes? Or could the good match between vision and bioluminescence spectra simply be a coincidence?

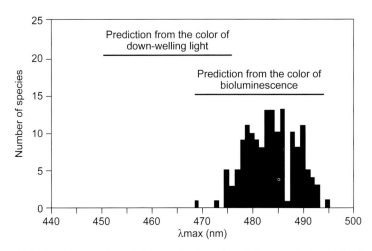

Figure 7.2. Distribution of visual λ_{max} of 175 species of deep-sea fish. The horizontal bars indicate the predicted ranges of λ_{max} values that would confer either maximum sensitivity to down-welling daylight, or maximum sensitivity to fish bioluminescence.

The spectral tuning of the visual pigments of the cottoid fishes of Lake Baikal in Eastern Siberia bears on this question. This lake is the largest (23,000 km³), the deepest (>1600 m, yet oxygen concentration never falls below 75–80 percent of the surface level), *and* the oldest (25–30 my) of all lakes in the world; the water is very clear. Some of its fauna, such as seals, have a direct phylogenetic relationship with marine fauna.

The spectral sensitivities of the visual pigments of the cottoid fishes, endemic to the lake, were studied as a function of the depth at which they were captured. A blue sensitivity shift was observed with depth, which was shown to result from four stepwise single amino acid substitutions in their rod opsin. The spectral sensitivity of the visual pigments peaks at ca. 516 nm in the littoral fishes, and at 484 nm in the deep-lake fishes. Yet no bioluminescence has ever been observed in Lake Baikal! Even more instructive, the light that penetrates to the depth at which the fish were caught is not blue, as in the oceans, but yellowish green, filtered by the yellowish upper layers of the lake. So the blue shift in visual sensitivity is not an adaptation to the spectrum of ambient light, and its interpretation remains an open question. While the study of visual pigments of Lake Baikal fishes cannot explain the evolution of sensitivity in deep-sea fishes, the more intriguing question is why there are not any bioluminescent organisms in the lake. Even bioluminescent bacteria have not been found, and a submarine search of the lake bottom, with all lights off, detected no luminescence. Why not? Answers can only be speculative, of course.

Going back to the oceans after this diversion, two groups of bioluminescent deep-sea fishes are particularly interesting, and fun to think of, anglerfish and dragonfish. There are some 18 families of anglerfishes and over 300 species, among them 11 families and 160 species of deep-sea anglerfishes (ceratioids, members of the order of teleost fishes called Lophiiformes). They spend most of their lives in the dark bathypelagic zone, below 2000 m. Females are characterized by a remarkable skeletal adaptation: their first dorsal spine, the illicium, is modified to carry—in different, humoristic ways—a lure full of bioluminescent bacteria (Figure 7.3). In some cases, this lure, called the esca, is stolidly planted right above the fish mouth, a bit as a lamppost; in others, it dangles at the end of a long and graceful line, many times the length of the fish. In all cases the purpose of

Figure 7.3. The anglerfish *Bufoceratias wedli*, ~5 cm long. The white "pearls" are freestanding lateral line organs that detect water vibrations.

Box 7.1. The Strait of Messina.

You don't have go to great depths to find deep-sea fish. Go to Messina, in Sicily, where you will find an array of bizarre fish in the local fish markets, a great many of them luminous. These come mostly from the waters near the Strait of Messina, where 30 miles from the strait the depth is more than 6000 feet.

You don't even have to go to the fish market; on the sandy beach to the north of the city, you can find hundreds of such fish freshly cast up as the tidal current sweeps through the narrow (3 km) and shallow (80 m) passage, which acts like a sill. But you must get there promptly; local cats know that the pickings are good. The two bodies of water have different masses, so their tidal amplitudes are out of phase, creating the tidal currents and permanent whirlpools, upwelling the deep water and fishes as it approaches the strait. The strait is featured in Greek and Roman mythology in numerous ways, the most notable being Scylla and Charybdis, sea monsters represented by rocks and whirlpools sited on opposite sides of the strait and greatly feared by sailors in antiquity. But

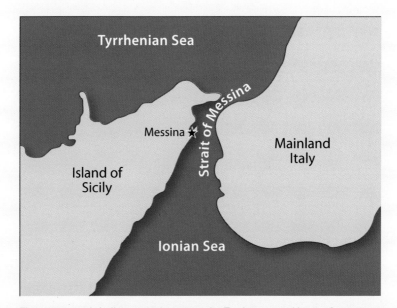

The narrow and shallow strait between the Tyrrhenian and Ionian Seas experiences swift tidal currents, bringing fish from the depths to the surface, where many are cast ashore.

there are relatively few studies of its bioluminescence. In the southern part of the strait, strong bioluminescence is localized at a depth of about 450 m at midday and moves upward in the evening. Some of this bioluminescence comes from photophores of fish, including *Argyropelecus hemigymnus* (a hatchet fish), *Cyclothone braueri* (a bristlemouth), and different myctophids (lantern fishes).

A constant glow in the water was attributed to luminous bacteria present in fish feces. Bacteria from the intestinal contents of several fish were examined; 100 percent were luminous, an apparently pure culture. Luminous bacteria may play the same role in fish intestines as *E. coli* does in those of mammals.

the luminescent lure must be the same, to attract swimming prey. Indeed, stomachs of anglerfish are typically found full of copepods, amphipods, fishes, and squids.

All anglerfishes share another outlandish trait, a most striking example of sexual dimorphism. The females can be almost spherical, flabby, and large (some species are 1 m long), with big mouths and strong teeth; presumably, they more or less stay put, content to wave their lure to attract food. The males, in contrast, are miniscule, usually less than 2 cm long (Figure 7.4). In some species, they attach permanently to the female, reduced to the life of sperm-producing machines. No male anglerfish emits light.

During their lifetime, anglerfishes migrate thousands of meters up and down. Adult females, at great depth, release eggs wrapped in a large amount of sponge-like material. Among the species where males are permanently attached to females, the male releases sperm upon a hormonal signal from the female. The sperm gets taken into this protective and buoyant material, where fertilization takes place. The fertilized eggs float up to the epipelagic zone, where they hatch and the larvae emerge. Metamorphosis then begins as the larvae descend in the water column. The males, destined to attach permanently to females, are endowed at first with large mouths, strong teeth, good eyes, and a strong sense of smell; they seek and then attach themselves to metamorphosed females at depths of about 2000 m, where the

Figure 7.4. The anglerfish *Linophryne bicornis* with both esca and barbel, as well as a parasitic male attached ventrally near the posterior.

females are most abundant, and a new life cycle begins. The attached males shrink in size, lose their eyes and teeth, and their mouth tissues fuse to those of the abdomen of the female; they acquire their nutrients through her blood and are essentially on board, nutritionally and physically, for the rest of their lives. Some females carry several such diminutive males—security in numbers!

Permanent attachment to females is not the universal rule among male anglerfishes. Some males remain unattached, while in some species, males bite and get a hold of a female only temporarily; little is known of such roving male behavior.

Going back to bioluminescence, the esca (the lure) of anglerfishes is mainly a chamber full of *Vibrio* bacteria. Bacteria can be isolated from the organ and grown outside the fish, but no culture has ever been found to emit light; the cultured bacteria may not have been those responsible for light emission. This chamber is closed except for an escape pore, which may also serve an entry for bacteria infecting newly metamorphosed fish. The fishes are able to control the emission, but it is not known how.

Most intriguing is the case of the ceratioids of the genus *Linophryne*. These anglerfishes not only sport an esca but also a luminous barbel under the lower jaw, together with a full set of huge teeth (Figure 7.5). Not all species of *Linophryne* exhibit as elaborate a barbel as *L. arborifera*; the barbel of *L. coronata*, for example, looks more like the esca on the top of its head. What is fascinating is that this second luminous organ of the *Linophrynes* does not harbor bioluminescent bacteria. In fact, the mechanism of barbel emission is unknown. Is it coelenterazine-based? No one knows. So far as is known, *Linophrynes* may utilize two different chemistries for emission.

Not a bit as whimsical as anglerfishes, but sharing the same waters and also endowed with a nasty set of teeth (Figure 7.6), three genera of dragonfishes are unique in emitting two colors of light, blue and red, from two different pairs of photophores. In addition to the blue-light-emitting photophores behind the eyes, shared by all dragonfishes, members of the genera *Malacosteus*, *Aristostomias*, and *Pachystomias* have a pair of large accessory photophores that emit red light, at ~705 nm, a color barely visible to the human eye (Figure 7.7). This red emission is apparently brought about either by filters or by a red-fluorescent screen, which modulates the blue bioluminescence emission; in either case, the emission yield of the primary bioluminescence emission would be reduced.

These blue- and red-emitting photophores can apparently be turned on or off independently. It is clear that to be able to signal to another fish of its species, or to illuminate its environment without being seen, would give a fish a clear advantage, provided that this fish could itself "see" red light. But how can it? What about their vision? While the visual pigment (rhodopsin) of most deep-sea fishes absorbs in the range 470–490 nm and is therefore sensitive to the blue emission of most bioluminescent fishes, the three red-emitting genera of dragonfishes have not one but two visual pigments, rhodopsin and porphyropsin (based on the same opsin), with peak sensitivities of both shifted towards the red, in the range 515–550 nm.

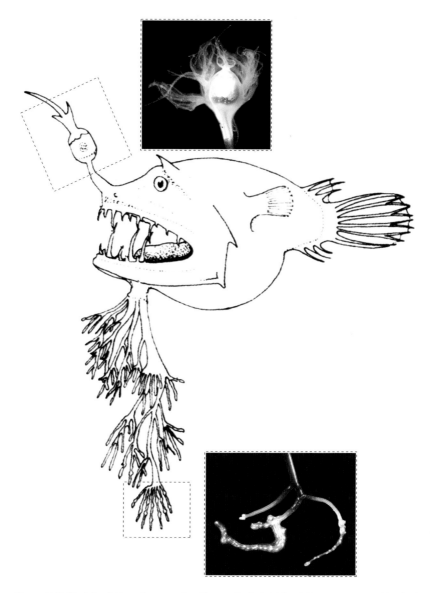

Figure 7.5. Sketch of *Linophryne arborifera* with its highly elaborate esca and barbel; the fish is about 5 cm long. The bioluminescent lure (esca) dangling above the fish mouth contains bioluminescent bacteria. The ventrally attached and highly branched barbel also emits light, but the mechanism of this bioluminescence is still unknown and does not involve bioluminescent bacteria.

Figure 7.6. *Photonectes waitti*, a recently discovered species of dragon fish from Samoan waters; in addition to its suborbital organ, next to the eye, it has bright pink lures in the middle and end of its barbel.

This is still hardly adequate for detecting true red light. A third pigment, with λ_{max} around 590 nm, was identified in the retina of *Aristostomias tittmanni* and *Pachystomias microdon*; it uses retinal as a chromophore coupled to a long-wavelength opsin. Although these three pigments would still be inadequate for detecting red light at 700 nm, it is *hypothesized* that these fishes may have a fourth visual pigment based on this new opsin and dehydroretinal, which would be expected to absorb maximally at ~670 nm.

In the retina of *Malacosteus niger*, on the other hand, pigments with peak absorbance at ~670 nm, presumably protein bound, were discovered and, surprisingly, identified as pheophorbides, derivatives of chlorophyll a. How could pigments absorbing red light bleach the pigments absorbing blue light, which are responsible for vision?

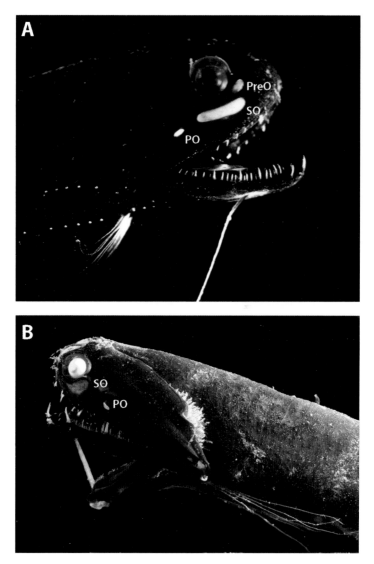

Figure 7.7. A. The head and jaw of a fresh specimen of *Pachystomias microdon* show-ing the positions of the suborbital (SO), the white postorbital (PO) that emits blue light, and the preorbital (PreO) photophores that emit red light. B. *Malacosteus niger* has only suborbital and postorbital photophores. Illumination by long-wave UV light shows the intense red fluorescence of the suborbital photophores.

This would require overcoming a large energy deficit. And how would a deep-sea fish acquire these chlorophyll derivatives?

The second question is less troubling than the first. *M. niger*, ag-gressive teeth notwithstanding, feeds mostly on large copepods, which feed on small copepods, which in turn eat photosynthetic organisms. Therefore, the fish could well acquire its chlorophyll-related visual sensitizers from its diet.

But the first question—how can low-energy photons of red light trigger the visual process, an energetically up-hill process?—has yet to receive a convincing explanation. Has it, in fact, been demonstrated that dragonfishes actually "see" their red light? Understandably, behavioral data have yet to establish that. At this time, the most attractive hypothesis is that the red-light emission attracts prey that do see it; red-light sensitive copepods? Stay tuned!

chapter eight

THE MANY FUNCTIONS OF
BIOLUMINESCENCE
Defense, Offense, Communication, and Propagation

Any reader reaching this point will have already concluded that the present-day functions of bioluminescence are many and diverse. Some seem obvious, even if not proven, and most have counterparts in devices or techniques used by man. For example, man used light signals to communicate long before it was known that fireflies do. Lanterns in the tower of Old North Church in Boston informed Paul Revere about the invading English at the start of the American Revolution: "One (lantern) if by land; two if by sea" in the celebrated Longfellow poem.

But consider camouflage. Adjusting the pattern and coloration of clothing or skin seen in reflected light is widely used by animals and man (chameleons, squid, or soldiers) for camouflage. Yet one technique used by animals but not by man (until very recently, perhaps) is to take advantage of light emission to obscure the silhouette. Seeing an airplane in the sky would not be possible if the plane itself emitted light from its underside with the same intensity and color as the sky above. Fish, squid, and other animals in the ocean do this, evidently to conceal themselves from predators below. It has actually been shown experimentally in several cases, including the small squid *Euprymna* (see Chapter 5), that the intensity of luminescence from the organism's underside is directly proportional to the incident intensity from above, which in nature would come from periods around twilight at dawn and dusk, moonlight near the surface, and sunlight deeper in the ocean. The animal gauges the intensity and, by various means in different animals, uses that information to adjust the emission from its light organ.

There are many other "tricks," which can be classed (with a small dose of anthropomorphism) under four major headings, each of

TABLE 8.1. THE FUNCTIONS OF BIOLUMINESCENCE

Category	Function	How achieved
Defense		
Deter, escape predators	Camouflage	Ventral emission, symbiosis
	Startle, frighten	Brief bright flashes
	Decoy, diversion	Luminous cloud, sacrificial lure
	Causes predators to avoid	Aposematism; danger signal
Offense		
Aid in predation	Startle	Brief, bright flashes
	Attract prey	Lure
	Aid in vision	See and capture prey
Communication	Courtship, mating	Flash signals
	Species recognition	Photophore patterns
Propagation	Enhance bacterial growth	Attract feeders

which may involve more than one specific type of function (Table 8.1). Altogether, one generalization that seems to apply in many cases is that a constant light emission attracts, whereas flashes deter.

The first category, *defense*, includes diverse functions that help the organism to deter and escape predators. Using light to startle or blind a would-be predator is very widespread. The coelenterates *Renilla* and *Aequorea* (Chapter 2) and the dinoflagellate *Lingulodinium* (Chapter 4) come to mind. To startle requires the element of surprise, so it must occur rapidly, without prior notice, and be brief, thus a flash of ~2 sec or less in duration.

Predators can also be avoided by a longer-lasting light emission. In complete darkness it would do an octopus no good to squirt ink, but ejecting a longer-lasting luminescence would confuse a predator and allow the luminous animal, which itself would then stop emitting light, to slip off undetected in the dark. This can be called the "squirt-and-run technique," and it might even be used with a brief flash. In the Sea of Japan, the crustacean *Vargula hilgendorfii* seems master of the technique (Chapter 1).

Scale worms (and some other marine annelids) use both flash and glow defensively. When attacked, they emit brief, repetitive flashes from the scales, which may startle or frighten the predator. If this fails to do the job, they shed the scales, which then glow continuously, as sacrificial lures to attract predators, while the animal itself swims away in the dark. Another polychaete worm, *Eusyllis*, goes further; it sacrifices its luminescing posterior part for the future well-being of its anterior part (Chapter 6).

Bioluminescence might also function in a defensive fashion as a so-called aposematic, or warning signal, and some experimental studies have provided evidence for this. Aposematism is the use of a readily identifiable (and memorable) external display by an organism to alert a would-be predator to a less-evident feature that would harm it; the rattle of the rattlesnake warns against its venomous bite, while some other snakes use bright coloration as a warning. A predator, having recognized from experience that the dangerous organism is an unfavorable prey, does not attack it. This is advantageous to both predator and prey.

The notion of such warning signals goes back to Darwin and Wallace, but only a few attempts have been made to determine if bioluminescence is used in this way. In one study, firefly larvae (glowworms), which emit a glow lasting tens of seconds or longer, were shown to be unpalatable to potential predators; this appears to be learned, for they subsequently avoid them.

Another example is the millipede *Motyxia* that lives on the floor of Sequoia forests in California (Figure 6.13). These animals emit a constant glow, albeit very dim, and when attacked release copious amounts of cyanide, from which they themselves are well protected. In a recent study, nonluminescent millipedes were attacked by a predator twice as often as luminescent millipedes. But no studies have tested the idea that predators actually learn to associate easy-to-see light emission with the cyanide to be avoided. Other nonluminescent millipedes also produce cyanide; these millipedes are often distinctively colored, another well-established aposomatic signal. The little limpet *Latia* (Figure 6.11) may be another example; its sticky luminescent slime can mark a predator in a way that its own predators can see, and the slime cannot be easily shaken off.

A second major category of functions of bioluminescence is *offense*, to aid in predation. In the simplest case, luminescence can provide light for an animal to see and capture prey. The flashlight fish is believed to take advantage of its two large light organs to see and feed in total darkness (Chapters 5 and 6), and it is very likely that lantern fishes and dragonfish also use their bioluminescence in this way. A more sophisticated example of using light offensively is that of the female *Photuris* firefly (dubbed the "femme fatale"); she mimics the flash code of a different species and attracts males for her dinner (Chapter 3).

Anglerfish use yet another strategy. They dangle a small luminous lure full of symbiotic bacteria near their mouth; its glow attracts the curious or the hungry, only for them to be seized by the fearsome jaws of the fish (Chapter 7).

There are at least two additional functional categories. One is *communication*, so elegantly illustrated in fireflies, which use it worldwide in courtship, but in different ways (Chapter 3). Species in the Northeast United States use a "simple" query and response sequence between individuals; but fireflies of Southeast Asia, as well as in regions of the United States, emit synchronously. While the latter mode certainly appears to involve communication, how it leads to mating remains a question. Another beautiful example is the cypridinid crustacean *Photeros annecohenae* in the Caribbean Sea, which uses an elaborate pattern of timed light pulses in courtship (Chapter 1).

While symbiotic luminous bacteria serve in all three of the above categories, depending on how the host uses the light, the light emission from bacteria may function in another way—namely, to enhance bacterial *propagation*. In many or most fish species, luminous bacteria are abundant members of the gut bacterial population and can thus be viewed as the *E. coli* of the ocean. So fecal pellets in the sea are very often luminous, and as these slowly fall to the bottom in the dark, their light emission attracts animals that feed on them, enhancing dispersal and propagation of the bacteria. At the same time, incidentally, feeding on such material contributes significantly to recycling of organic carbon in the ocean, which is of special importance below the euphotic zone, where if not for luminescence it would be effectively unavailable as food. Dead animals, ranging from fish to small crustaceans, add to the luminous detritus, which is selectively colonized by luminous bacteria as a first step in its decomposition.

The light emitted by luminous mushrooms may function similarly by attracting insects; they eat spores, which pass through their digestive tracts unharmed and are thereby dispersed.

chapter nine

THE ORIGINS AND EVOLUTION OF BIOLUMINESCENCE

How Did Luciferases Originate?

There are many different luciferin/luciferase systems, widely distributed phylogenetically. Why, when, and how did this happen? An exciting hypothesis for the origin of luciferase genes was first proposed some 50 years ago, and although not pursued at the time, it has been revived in various forms since. It was based on the realization that oxygen, which did not appear in significant amounts until about 2400 Mya (million years ago), well after primitive life was already established, would have been toxic for cells. Thus, luciferases and their luciferin substrates, with the ability to utilize oxygen and reduce its level in cells, could have evolved in response to the appearance of oxygen. This might have occurred independently many different times.

The earth was formed about 4600 Mya but lacked oxygen in the atmosphere for about the first 2 billion years. Evidence indicates that life originated before 3400 Mya, sustained for the next billion years or so by harnessing energy from the sun (anaerobic photoautotrophs) and from inorganic chemical reactions (chemoautotrophs), as well as by heterotrophic metabolism, as found today in many anaerobic microbes.

Bioluminescence could not have originated during this first period of life on earth (3400–2400 Mya), since oxygen is required for all bioluminescent reactions. Primitive life, having been in existence for about a billion years in the absence of oxygen, would not have easily accommodated to the ravages this newly arrived oxidant might wreak on cells.

The accumulation of oxygen in the atmosphere was first detected in the geological record at about 2400 Mya and attributed to an ancestral cyanobacterium that evolved oxygenic photosynthesis, capturing energy from the sun and in the process fixing carbon and producing oxygen. While this process was the beginning of what has been called the "great oxidation event" (GOE), the oxygen levels remained relatively low for about 1600 million years, when a second and larger increase in oxygen started. Between about 800 and 400 Mya, oxygen levels reached half or more of present-day values (Figure 9.1). This period coincides with the Cambrian Explosion, leading to a great proliferation and diversification of living organisms.

The actual oxygen levels reached in the time between the GOE and the Cambrian are not well established, and they were surely not constant over that time nor evenly distributed in the oceans, where life was evolving. Estimates of atmospheric levels put it at from less than 1 percent of present atmospheric levels (21 percent O_2) to only a few percent. Levels in the oceans could have been much lower.

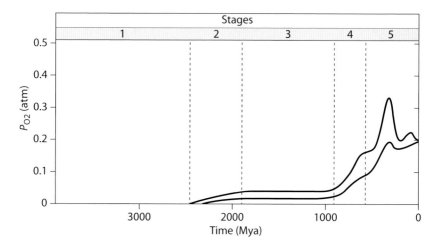

Figure 9.1. The concentration of oxygen along evolutionary time. During stage 1, O_2 was essentially absent from the atmosphere and oceans. During stage 2, oxygen began to pervade the atmosphere and surface ocean, but levels remained at 1–10% of modern levels and large volumes of subsurface ocean water remained oxygen-free. Only during stage 4 did oxygen begin to approach modern abundances. Stage 5—the Phanerozoic, or the age of visible animal life—is the one interval of Earth history when oxygen was generally plentiful in the atmosphere and throughout the oceans. The two traces represent upper and lower estimates.

In the long run, the introduction of oxygen turned out to be a major advance for life on earth. But at the time it might have been a near-disaster; oxygen and its several active chemical forms (e.g., hydrogen peroxide, singlet oxygen, superoxide, oxygen radicals) can be toxic for cells; these active forms of oxygen react vigorously and destroy other molecules, including biomolecules. The repeated admonitions we get to include antioxidants (e.g., vitamins C and E) in our diet may be excessive, even harmful, but they attest to their believed importance. That high oxygen can be very damaging was sadly discovered only after the development of retinal detachment and blindness in numerous premature infants who had been kept in tents with 100 percent oxygen.

Living organisms nevertheless survived, and they likely did so by evolving ways to tolerate oxygen and escape its destructive effects, and at the same time benefit from its rich reward—respiration—used by most present-day organisms to obtain energy from fixed carbon. While the fermentation of a sugar molecule in the absence of oxygen yields two ATP molecules (the energy currency in biological systems; see Glossary), respiration with oxygen yields about fifteen times more. Respiration made possible the explosion of life in the Cambrian and the evolution of large animals with highly efficient methods for transporting and providing oxygen to cells and tissues throughout the body.

How did primitive organisms survive the new threat from the toxic oxygen? The hypothesis that luciferases originated as a way to remove oxygen—the light itself being a nonfunctional by-product—is revisited here with greater knowledge of the early earth, cell biology, and bioluminescence than 50 years ago. All the different forms of life extant at that time would have been challenged. Those that evolved a luciferin/luciferase system as a solution would likely have done so independently, and the chemistry of each might have been different; the only common feature would have been a reaction that consumed oxygen avidly and nondestructively. The fact that all present-day luminous organisms require oxygen, yet may differ widely in the nature of their luciferases and luciferins, finds an explanation in this theory.

Would a luciferase reaction have been able to remove enough oxygen? When oxygen first appeared, it presumably could have done so, since the rise in oxygen must have occurred slowly over time, perhaps over many hundred millions of years. So initially the amount of oxygen that had to be removed would have been small, and with time more luciferase could have been produced to keep up with the increases in oxygen. In present-day luminous bacteria, for example, at least 5 percent of the soluble protein is luciferase, and the reaction accounts for about 20 percent of their oxygen consumption.

Other methods to combat the destructive effects of oxygen most certainly evolved concurrently with luciferases, and as the atmosphere came to have higher levels of oxygen, these could have ensured the survival of cells as luciferases became inadequate to do so. The luciferin/luciferase reactions would then have been lost by selection unless cells evolved some function with survival value for light emission itself, as in present-day systems. For example, juvenile squid actively select luminous bacteria to populate their light organ; nonluminous mutants are rejected even in the absence of a luminous alternative (see Chapter 5). This means that when this occurred, systems were already in place to detect the emitted light. Remarkably, in contrast to the chemical diversity of light-emitting mechanisms, the light-detecting systems in animal vision have been basically conserved throughout evolution. All are based on the light induced cis–trans isomerization of a small molecule (i.e., retinal) coupled to a protein (opsin). In this scenario, it is assumed that anaerobic bacteria, which had been present for millions of years by the time of the GOE (~2400 Mya), were the first cells to be challenged by oxygen. The properties of luminous bacteria today support the view that its luciferase evolved when oxygen levels were very low. It has an extremely high affinity for oxygen, removing it efficiently at ambient oxygen concentrations 1000 times lower than today's atmospheric level, and therefore the probable oxygen concentration at the time it evolved.

Experiments have also shown that if present-day luminous bacteria are grown in pure oxygen, they produce more luciferase and more enzymes, such as superoxide dismutase, that will destroy toxic forms of oxygen. This indicates that oxygen detoxification mechanisms remain responsive to oxygen concentration. It is also significant that respiration (growth in oxygen) requires a much higher oxygen concentration than luminescence does. Also, luminous bacteria grow well and produce luciferase in very low oxygen, even in its complete absence. Upon the admission of oxygen, luminescence occurs immediately.

But what about more complex organisms, such as jellyfish, crustaceans, and fireflies? They did not obtain their luciferases from ancestors of present-day luminous bacteria; their luciferins and luciferases are different, as are their luciferase genes. Today, their luciferases are retained because their bioluminescence has survival value; the functions for light emission itself may have evolved as oxygen levels increased and the role of luciferase in detoxification became inadequate or redundant. Might luciferases in higher eukaryotic organisms have also originated as a way to lower cellular oxygen, and only later been coopted for the functional value of the light? What can be said along these lines?

One fact that contributes to the evidence for an early origin of firefly luciferase is that, like bacterial luciferase, it has a very high affinity for oxygen; this again suggests that it first evolved at a time when environmental oxygen was very low. Moreover, also like luminous bacteria, luminous beetles exposed for a day or so to pure oxygen produce greater amounts of luciferase, as well as catalase and superoxide dismutase, alternative ways to remove other forms of oxygen. This also suggests that beetle luciferase was selected originally for its oxygen utilization rather than for its light emission.

An intriguing fact is that firefly luciferase is located in the peroxisomes (see Chapter 3), organelles in which there are many other enzymes that utilize molecular oxygen. Peroxisomes occur in cells of all higher organisms, indicative of an early origin; with their abundance of oxidative enzymes, they themselves have been proposed to have evolved as a way to reduce oxygen levels in early cells. The possibility that peroxisomes themselves might have originated as bacterial endosymbionts, and so could have brought with them a luciferase that evolved independently of other luciferases in a separate bacterial line, was proposed but later disputed. It is nevertheless well established that genes can be mobile, so the luciferase could have established residence in peroxisomes in some other way.

In the course of evolution, a gene could be duplicated, and a mutant of the copy selected for a new function, unrelated to the original one. As a result, the two genes would be related at the level of sequence but not function. Such a situation occurs in two bioluminescence systems, firefly and jellyfish; their present-day luciferases have been found to have similarities to functionally unrelated genes.

Firefly luciferase is similar in sequence (homologous) to long-chain acetyl CoA synthetase (ACoAS), an enzyme central in sugar metabolism in both prokaryotic and eukaryotic cells, whose gene occurs widely in virtually all aerobic organisms, from bacteria to animal cells, suggesting that it evolved at an early time. Although it has been assumed that firefly luciferase evolved from ACoAS, it seems equally or more plausible to conclude that ACoAS evolved from an early firefly luciferase and that firefly luciferase was one of the survivors for which a function of light emission was discovered in evolution.

Jellyfish (coelenterate) luciferase (apoAequorin) is also homologous to another protein, calmodulin, a calcium-binding protein that mediates many cellular processes. Here again, it can be postulated that calmodulin evolved very early from a jellyfish-type luciferase, not the other way around. It is of interest that the luciferin in this reaction, coelenterazine, is a potent antioxidant that occurs widely in nonluminous marine organisms.

Living organisms today live on the brink, constantly threatened by the possibility of oxidative damage but armed with an array of antioxidant defense mechanisms. But at the time oxygen first appeared on earth, life may have been in an even more precarious situation. Did it survive thanks to bioluminescence? Do we owe our existence to luciferases? Not exclusively, to be sure, for other mechanisms and antioxidant molecules may have evolved concurrently. So far as we know, luciferase systems of today's organisms do not have an antioxidant role; they are retained only because the light emitted endows the organism with a selective advantage.

The "oxygen detoxification" hypothesis is of course only a hypothesis; we do not really know. However, we can consider how, if valid, it might account for several puzzling aspects of luciferase systems.

1. *There are many different genes and corresponding biochemical mechanisms in different luminous groups. How could this have occurred?* On the assumption that when oxygen began to appear the extant living organisms were broadly distributed in the earth's oceans, all would have been subjected at the same time to the same evolutionary pressures and would have generated independently many different genes, such as we find in those that have survived to this day. This is convergent evolution. There may have been other luciferases, maybe many others, that evolved but were lost without leaving a trace.

2. *Bioluminescence is found in some 50 percent of all phyla, and in 35 percent of all classes within those phyla; however, considering individual species, the percentage of luminous taxons is vanishingly small. Why?* Speciation has had a billion years or more to work, but the niches where luminescence has a selective advantage are relatively few. So for the many new species that evolved from an original luminous line, selection was based on something other than bioluminescence.

3. *Bioluminescence occurs predominantly in marine organisms. Why?* The evolutionary events postulated to have given rise to luciferase systems occurred when all living organisms were oceanic. The transitions to systems in which selection was based on the function of the light itself would also have occurred mostly before life on land and the Cambrian Explosion; in most cases, the function of luminescence that luminous pioneers

brought with them may not have been readily transfer-
able to the new terrestrial environment and was lost.

In concluding, we want to emphasize that we have drawn evidence for the oxygen detoxification theory from only three extant
luciferases—bacteria, fireflies, and jellyfish—and that the evidence is
suggestive but certainly not conclusive. The origin and evolutionary
path for other luciferases, of which there are many, may have been
different.

For most of the other luciferases there is little evidence to go
on. But for two groups, the fungi and the cypridinids, the oxygen
requirements of their bioluminescent systems are known to be high,
indicative of an origin at levels corresponding to present-day values,
thus unlike the bacteria and fireflies. So we must conclude that even
if some luciferases originated as a mechanism for oxygen detoxification, the evidence indicates that this was not so for all.

part three

BOOKENDS

The two chapters of Part III play the role of bookends. Chapter 10 samples the many applications of bioluminescence that have been made possible from our understanding of its mechanisms in different biological systems and advances in molecular biology. Chapter 11 starts with short course on how light interacts with atoms and molecules, and then shows the reader how some chemical and biochemical reactions result in the emission of light.

In Chapter 10, we will limit ourselves to only a few examples. Choosing these examples was a difficult task, as there are now so many to pick from. Every day brings new papers in which GFP or other fluorescent proteins—the poster children of bioluminescence research—are the critical tools, and it is clear that medical applications of bioluminescence with one of the several luciferases can be expected to become more and more important in a world shying away from radioactivity.

Everyone knows that light causes chemical changes; witness dark glasses and sunscreens. But few of us know how this happens, and even fewer realize that the photochemical processes that result in chemical changes can work the other way around, with the emission of light as the outcome. Chapter 11 is a very short course in photochemistry and a guided tour of the field of chemi- and bioluminescence based on our present understanding, still sketchy, of these phenomena.

Figure 10.1. Children catching fireflies at a festival in Japan.

chapter ten

APPLICATIONS
Tools for Biology, Medicine, and Public Health

Carolus Vintimillia of Sicily, who first proposed, in 1647, that fireflies flash for sex, would be surprised to hear that it is now done for money–big money. Today, several important applications of bioluminescence are now commercialized. But many, many fireflies, collected by children for pennies (Figure 10.1) had to be sacrificed to carry out the basic research needed to demonstrate the capability of their luciferase/luciferin to generate light in the presence of ATP, thus making available a most sensitive test for ATP. Along parallel lines, the arduous work of deciphering the light emission pathway of coelenterates, which similarly required innumerable jellyfish, led to a very sensitive test for free calcium in cells.

The power of these assays was immediately recognized. But the full scope of bioluminescence capabilities had to wait for the methods of molecular biology, which made the various different luciferase genes widely available and allowed their insertion, practically at will, into any DNA sequence. It is not an exaggeration to say that basic work on bioluminescence has now transformed the panorama of biological research.

The fluorescent protein GFP, discovered as a part of the jellyfish luciferase system, started its own revolution. Inserted as a tag in a DNA sequence, GFPs reveal by fluorescence the role in space and time of whatever mechanism is under investigation; genes of luciferases can be similarly manipulated to result in a bioluminescence signal.

Not long ago, bioluminescence was viewed mostly as a fascinating feature of the living world. It was put in the category of "basic research," knowledge for the sake of knowledge, exploration for the fun of it, with no particular practical use in mind. The study of its

genetic and chemical basis had seemed a most unlikely area of research to contribute in any practical way. Now, however, bioluminescence applications are numerous, and while the field is too vast to be
fully treated here, we will provide a few examples, showing how un-
predictable the pathways to new knowledge can be.

Early Applications

Practical application of firefly bioluminescence emerged as soon as
its ATP requirement for light emission was discovered (Chapter 3);
the amount of light emitted is directly proportional to the amount of
ATP consumed. Such assays were shown to have three major advan-
tages over others: (1) rapidity (only minutes are needed); (2) sensitivity (a billion times less ATP can be detected by luminescence than by
conventional methods); and (3) proportionality over a vast concentration range.

Assays were quickly adopted for the measurement of ATP in research and are now done on a large commercial scale for several different practical applications. All living cells contain ATP; therefore
assays that detect ATP can immediately tell of contamination—for
example, by bacteria or mold. Carcasses in slaughterhouses can be
checked for fecal (bacterial) contaminants within minutes; tradi-
tional culturing methods take a day or so, long after the product has
been processed and shipped.

In high-sugar soft drinks, slow-growing mold is a potential though
not frequent contaminant, but it is very costly in dollars and brand
reputation if a contaminated product reaches the consumer. Automated firefly luciferase/luciferin through-flow assays have been developed to check for mold; all batches can readily be tested. Other
examples of the applications of the firefly system to detect undesir-
able microbes would make a long list.

Today, large amounts of firefly luciferase are used in commercial
applications, but it is no longer extracted from fireflies collected by
children as in the days of discovery. It is now synthesized by recom-

binant DNA techniques, and firefly luciferin is also a chemically synthesized laboratory product. A widely used DNA sequencing technique, called pyrosequencing, is now by itself a major consumer of both luciferase and luciferin, and has contributed to the great reduction in the time and cost of sequencing DNA.

The procedure starts with a single DNA strand to be sequenced. By following the synthesis of the complementary strand, nucleotide by nucleotide, it can be determined which of the four nucleotides reacts and is thus complementary to that position in the single strand. One of the four is added to the assay; if it reacts in the synthesis step, a molecule of pyrophosphate is produced, which is then converted to ATP, causing a flash of light by reacting with firefly luciferase and luciferin in the assay. If no flash occurs, the nucleotide that was used is not the correct one; the next nucleotide is checked until the right one for that position is identified. The procedure is then repeated with the synthesis of the subsequent position in the unknown DNA sequence.

While we owe to firefly bioluminescence a test for ATP, the very sensitive determination of intracellular calcium concentration has relied for years on the jellyfish (*Aequorea*) system and its calcium-regulated protein aequorin, which is a stable enzyme peroxidic intermediate of the coelenterazine chromophore (Chapter 2). The presence of calcium triggers its immediate decomposition with emission of a flash of light; the peak amplitude of the flash is proportional to the calcium concentration. Because of aequorin's high sensitivity (down to 10^{-7} M) and fast response time, it was for years the probe of choice in physiology. It required intracellular injections, however, which can only be easily managed in large cells.

The Molecular Biology Era

Making it possible to produce large amounts of a given protein once its sequence has been deciphered is evidently a great asset. But another important gift of molecular biology techniques is the ability to insert a gene of interest in a specific stretch of DNA and to see the consequences of this insertion. Because of the uniqueness of light as a signal, as well as the sensitivity and fast response of light measurements, bioluminescence reporter genes are widely used to monitor biological processes. Using genetic engineering techniques, biotech companies are synthesizing various different luciferases and developing assays. The DNA sequence of firefly luciferase, for example, can be used as a "reporter" of the activating or suppressing effect of a given promoter in a DNA sequence. Often, a second and different luciferase, such as *Renilla* luciferase, emitting at a different wavelength, is introduced in the same cells as an internal standard or to track a different process.

Not only can the DNA that codes for a luciferase be inserted in a given cell, but it can also be attached to the control region of a target gene responsible for the production of a different substance of interest. Thus, when the target gene is expressed in the cell, the luciferase will also be produced; when and where it is produced can be determined by light emission, all in a noninvasive way—one has only to observe. For example, the on and off expression of a specific gene can be tracked over the course of time, whether it occurs during the day or the night phase, or changes during development, for example.

It's well known that circadian rhythms—rhythms with a ~24 hr period—govern the expression of many genes in relation to time of day; the bioluminescence of dinoflagellates provides us with a spectacular example (see Figure 4.10). Until about 1988 the belief was that only eukaryotes had circadian clocks. Scientists then had the idea of inserting the luciferase genes of bacterial bioluminescence (Chapter 5) as reporters in the genome of the prokaryote cyanobacterium *Synechococcus*, where hints of a 24 hr rhythm had previously been reported. The results were spectacular: in the presence of a long chain aldehyde, necessary for bacterial emission, the cyanobacteria emitted bioluminescence with daily periodicity (Figure 10.2).

This is a striking demonstration of the power of bioluminescence as a molecular biology tool. Emission intensity can be easily moni-

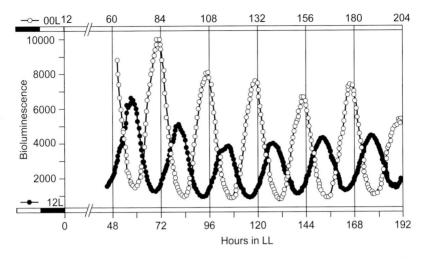

Figure 10.2. Circadian rhythm of bioluminescence emission from cyanobacteria. The bacteria were cultured under a 12 hr/12 hr light/dark (LD) cycle, then transferred to vials for measurement of light emission in constant dim light (LL). The two traces were from cultures grown on light-dark cycles 12 hrs out of phase.

tored and recorded, bacterial mutants can be equally easily generated and detected, and the search for mutants of different periodicity automated. In the nearly 20 years since the original observations, random insertions of the luciferase genes in *Synechococcus* revealed expression rhythms throughout its genome.

Another example is the use of bioluminescent imaging to locate tumors in living bodies, using ultrasensitive cameras and techniques based on antibodies. One recent example, chosen among many, is based on the cypridinid luciferin/luciferase system (Chapter 1), not yet mentioned in this survey of bioluminescence applications. Its goal was to create a bioluminescent probe that emits red light, which is absorbed less by blood than blue light and so therefore travels farther through tissues. Such a probe would be used to tag a specific protein on the surface of cancer cells. Cypridinid luciferase is a stable, secreted protein with a high turnover rate and two glycol chains to which, in the work discussed here, a far-red emitting indocyanine dye was attached (Figure 10.3).

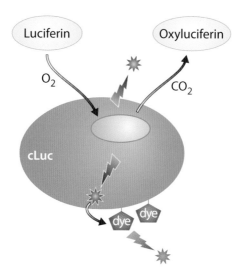

Figure 10.3. Cypridinid (*Vargula*) luciferase (cLuc; large green oval) catalyzes the re-action of Cypridinid luciferin with oxygen at its active site (small oval) to give oxylucif-erin and blue light (blue star). However, if molecules of an indocyanine dye (red penta-gons) are coupled to the two glycol chains of luciferase, then energy transfer to the dye may occur before emission of blue light; in this case the emission is red (red star). Red light transmits better than blue light through tissues and blood, an advantage in imaging.

Here Bioluminescence Resonance Energy Transfer (BRET) comes into play. The name BRET refers to the mechanism of the transfer of energy from the electronically excited product of the enzymatic reaction to the dye (see Chapter 11). The energy donor, electroni-cally excited oxidized luciferin, passes on its excitation energy to in-docyanine, a transfer made efficient by their both being attached to luciferase at the time of the reaction with oxygen. As a result, the enzymatic oxidation of cypridinid luciferin by this modified oxylu-ciferin produces a bimodal spectrum, with a peak at 460 nm pro-duced by excited oxyluciferin itself and a second peak at 675 nm emitted by the dye. This 675 nm peak becomes dominant in the body because blood absorbs the 460 nm emission. The luciferase conjugated to the dye is then linked to an antibody raised against antigens present on the surface of many cancer cells. Experiments with rats showed the potential of the method.

The Fluorescent Proteins

Bioluminescence can be a tool for assessing the role of distance between molecules in a biological system. The usefulness of the modified cypridinid luciferase discussed above is based on the proximity of the luciferin and the dye on the luciferase (Figure 10.3). Many examples take advantage of fluorescent proteins such as GFP, or of more recently discovered fluorescent proteins from corals. These coral proteins share with GFP the same general "β-can" crystal structure (see Figure 2.5) protecting the fluorophore, itself a close relative of the GFP's fluorophore.

Nature gave us the best examples of BRET in *Renilla* and *Aequorea* (Chapter 2). *In vitro*, the luciferases of these two coelenterates catalyze the oxidation of coelenterazine by oxygen with emission of blue light. But in the intact animals, the emissions are green, because of the presence of GFPs in very close physical association with the respective luciferases/luciferins. These two GFPs have the same fluorophore–three modified amino acids in the protein chain–but the amino acid sequences of the rest of the protein in the two GFPs are very different, even though their barrel-shaped structures are very similar. Over the years, a true rainbow of fluorescent proteins, from blue to red, has become available through mutations of the amino acid chain of GFP. These have become invaluable tools in biomedical research, enabling different animals or even cells to be labeled by a unique color, allowing them to be distinguished from all others in an experiment. As shown in Figure 10.4, for example, two variants of GFP, a yellow-green (f) and a cyan (g), introduced in a stem cell line stably label mouse embryos, providing sources of tagged cells. This labeling persists in the adult stage.

The availability of fluorescent proteins of different colors has a special impact in neurobiology, where an important question is how connections between nerves are established—thus how learning occurs and perhaps how memories are laid down. A goal is to understand how, as the brain grows and is remodeled from embryonic life through childhood and adulthood, its neural circuitry is established and functions.

The many dendrites of a mature nerve cell receive synaptic inputs from many different axons, thereby integrating information. To study how new synapses are established in a live zebra fish larva, growing dendrites of one neuron were labeled with a red fluorescent coral protein, while the new synapses, labeled with GFP, appear as green dots. The cell was filmed for 10 days (Figure 10.5). The results show that

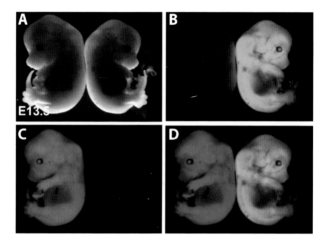

Figure 10.4. Dual noninvasive reporter visualization in mouse embryos expressing GFP variants. *A.* Image in white light, no epifluorescence. *B, C, and D.* Images taken through filters allowing visualization of YFP, CFP, and both, respectively.

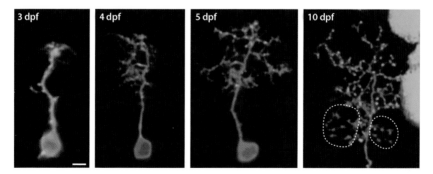

Figure 10.5. Formation of new synapses in a zebra fish larva. The growing dendrite images of individual cells on (*left to right*) days 3, 4, 5, and 10 post-fertilization (d.p.f.) show in red dendrite growth and its arborization, with filipodia that seek and form synapses (*green*) with axon candidates over the several days. Some of the new synapses are abandoned, as seen in the 10 d.p.f. image (*no green spots within the dotted circles*).

the formation of new synapses occurs on a trial basis. Those not of functional value, presumably, do not last long and are abandoned; others are accepted as the network develops. The longer-term challenge is to understand the basis for acceptance or rejection of nascent synapses in individual cases. There is evidence in support of the long-held idea that pathway use serves in synaptic establishment and stabilization.

Mapping neuronal circuits, with the long-range goal of understanding how nerve connections account for brain functions, is perhaps the place where fluorescent proteins show their most exciting promise, albeit in the distant future. For such studies, "Brainbow" fluorescent proteins, which can be engineered to label individual neurons in the brain and thereby track pathways and identify synaptic

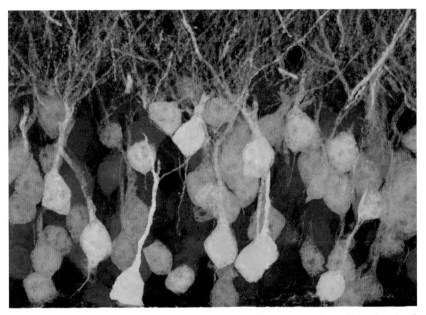

Figure 10.6. Using a confocal microscope, it is possible to image the cellular details of the dentate gyrus of the hippocampus in a "Brainbow" transgenic mouse. Each nerve cell is revealed by one color of the rainbow palette produced by the combinatorial expression of three different colored fluorescent proteins. Using the tools of genetic recombination, Harvard scientists have induced individual neurons to express red, green, and blue fluorescent proteins in random mixtures that make it possible to see each individual cell and its branching processes.

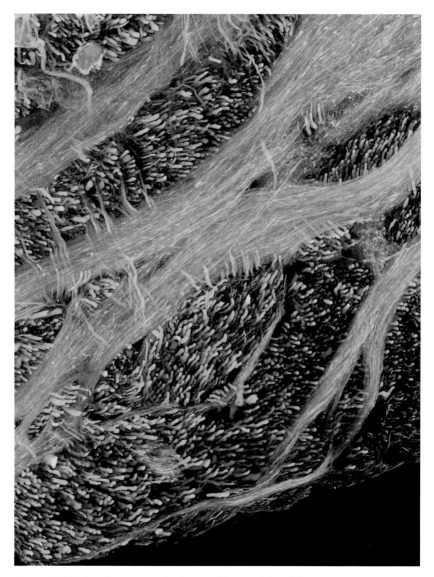

Figure 10.7. Using the Brainbow technique, it is possible to follow nerve cell processes long distances. In this image the long axons of nerve cells in the mouse auditory system are easily visible. The color labeling allows axons to be traced from one place to another, helping scientists to understand how auditory information is organized in the brain.

connections, are now available to the experimentalist on a vast pallet (Figure 10.6). It is certainly one place where art meets science.

Bundles of brainbow-labeled axons in a pathway are reminiscent of telephone cables before fiber optics, in which each of dozens of strands had a unique color-coded covering (Figure 10.7).

chapter eleven

HOW DOES LIFE MAKE LIGHT?
"Excited Molecules" and Bioluminescence

Dyes may fade in the sun, because light and air alters their chemical structures. Conversely, chemical reactions can produce light in a process known as chemiluminescence, or as bioluminescence if a living organism participates in the process. The challenge is to identify, in a bioluminescent organism, not only the specific molecular species that emit the photons but also the chemical and enzymatic processes by which such molecules acquire the energy needed for emission. A clue to our understanding of bioluminescence is that photons of blue, green, or yellow light carry amounts of energy comparable to those that bind atoms in molecules.

For a molecule to emit light, it has to be in an "excited" state. Just as energy is required to throw a ball into a high basket and after a brief pause most or all of this energy is released when the ball falls to the ground, in bioluminescence it's the energy given up by a chemical reaction that puts a molecule in a high basket, an "excited state." When the molecule falls back to its "ground state," energy is released, sometimes as light, but more often as heat.

How Does Molecular Excitation Work?

Molecules are geometrically well-defined assemblies of positively charged atomic nuclei surrounded by clouds of negatively charged electrons. The electrons are organized in compartments of space of different shapes and energies about the nuclear framework. Electrons "occupy" these so-called molecular orbitals, filling them in order of low to high energy. Only two electrons are allowed in each orbital. Some electrons are very close to only one nucleus or to a pair of nuclei; others are "delocalized" over the entire molecule.

Each electron also spins, top-like, in one direction or the other. The two electrons in an orbital are said to be spin-paired if they spin in opposite directions. If all electrons in a molecule (containing an even number of electrons) have paired spins, the molecule is said to be in a "singlet state." If, however, two orbitals in a molecule are each occupied by only one electron, and if the two electrons spin in the same direction, the molecule is in a "triplet state." Triplet states are of lower energy than singlet states of the same electronic configuration. The distinction between singlet and triplet states is important with respect to both reactivity and energy-dissipating properties of a molecule. When the electrons occupy the *lowest* energy set of orbitals (where the electrons are spin-paired) the molecule is in its electronic *ground state* (in this case, singlet state). If, instead, one or more higher-energy orbitals are occupied, the molecule is in an electronic *excited state*. Apart from rare exceptions, such as in the all-important oxygen molecule (to which we will return), all electrons are spin-paired in the ground state of molecules. In addition, individual atoms and groups of atoms within a molecule vibrate, and the molecules as a whole rotate, with all those motions obeying rigorous laws.

Light may be emitted by a molecule if one of its outermost electrons, orbiting furthest away from the nucleus, is separated from its partner and "promoted" from its ground state level, called S_0, to a higher energy level. In the left part of Figure 11.1, which only shows the ground state, S_0, and the first excited state, S_1 (to simplify the figure, higher excited states are not shown), these two electronic states are represented by heavy horizontal lines; thinner lines represent the associated levels of vibrational and rotational energies. The molecule may be promoted to an excited state by the absorption of photons, represented by the vertical arrows, by a chemical reaction, or by very high temperature.

Independent of the mode of excitation, the processes by which such excited molecules lose their surplus energy and fall back to the ground state are the same. They occur extremely quickly from excited states higher than S_1 (such as S_2 or S_3, not shown in Figure

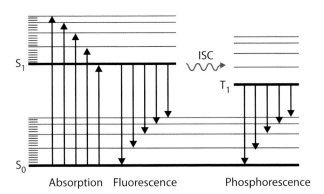

Figure 11.1. Transitions between molecular energy levels (so-called Jablonski Diagram; to simplify, only the first excited state, S_1, is represented, not S_2, S_3, etc.). In the ground state all electrons are paired, with rare exceptions, and the molecule as a whole is in its singlet (S_0) ground state. Following *absorption* of a photon (a very fast process, $\sim 10^{-15}$ sec, represented by an up-pointing vertical arrow), an outermost electron is promoted to the higher electronic energy level S_1 in one of its associated vibrational and rotational energy levels, with no change in the direction of its spin; the figure shows several such arrows, as expected if the molecule is exposed to "white light" and thus to photons of different energies. Excess vibrational or rotational energy is dissipated fast via collisions with other molecules (in $\sim 10^{-13}$ sec, compared to the 10^{-8} or 10^{-9} sec lifetime of the S_1 state) before emission of a photon of *fluorescence*. This brings the molecule back to its ground state, S_0, usually with an excess of vibrational and rotation energy. In some cases, determined by the structure of the molecule, a process called *intersystem crossing* (ISC), may change the direction of spin of the promoted electron. This brings the molecule from S_1 to its triplet state, T_1, from which it may slowly emit *phosphorescence*.

11.1) to S_1. Collisions with other molecules then remove the excess vibrational and rotational energy associated with S_1. Within nanoseconds to microseconds (10^{-9} to 10^{-6} sec), the excited molecule may then fall back from S_1 to the ground state S_0 by emitting a photon of fluorescence.

Occasionally the direction of the spin of the electron that had been promoted to an excited state may be inverted, by a process called intersystem crossing (ISC), and the molecule now finds itself in a triplet state (right side of Figure 11.1). De-excitation requires a second switch of spin direction, a relatively slow process, causing triplet states to have longer lifetimes than singlet excited states and, consequently, to be much more susceptible to de-excitation by collisions with other molecules, especially oxygen. Emission from a triplet state is called phosphorescence. Both fluorescence and phosphorescence are types of luminescences, as are chemi- and bioluminescences.

Looking at the case of diatomic molecules is instructive (Figure 11.2). The fluorescence spectrum (and therefore the fluorescence color) of a molecule is shifted to the red of its absorption spectrum because some of the light energy initially absorbed is dissipated as heat before fluorescence is emitted, and because after emission the molecule is left with a surplus of vibrational energy.

Applied to the more complex case of polyatomic molecules, the same basic principles explain why fluorescence spectra are red-shifted from absorption spectra. They tend to be mirror images of one another, because the geometries of the molecules are essentially the same in the ground and excited states. This is illustrated by the spectra of tetracene, a beautifully symmetrical and intensely fluorescent molecule (Figure 11.3).

Occasionally, an excited molecule can donate its excitation energy to another, different molecule. This can evidently happen in a trivial way, if the light emitted by one molecule is absorbed by another, which becomes excited in the process and, in turn, radiates this newly acquired energy. This process depends, obviously, on how closely the absorption spectrum of the "acceptor" molecule overlaps

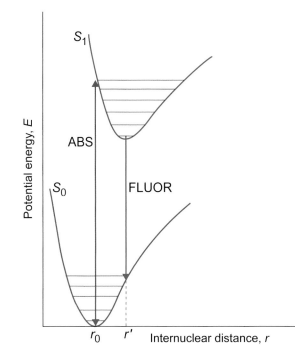

Figure 11.2. The potential energy E (y-axis) of a diatomic molecule as a function of the internuclei distance, r (x-axis), is represented by a blue curve in the ground state. If the nuclei get closer than their equilibrium distance, r_0, E increases because the nuclei re-pulse each other, while if they get further apart, the energy also increases because the bond between the atoms is stretched (or even broken). The bond is weaker in the electronically excited state, and as a result the corresponding red curve is altogether displaced to the right from the ground state curve. The positions of the nuclei cannot change during the very fast processes of absorption or emission of a photon; only electrons have time to change their positions (Franck-Condon Principle). The blue and red vertical arrows between ground and excited state represent the most intense tran-sitions between the two states. The blue vertical arrow hits the red upper curve at a point higher than its minimum (at the equilibrium distance r' between the nuclei in the excited state); this causes the molecule to vibrate back and forth, thousands of times, to dissipate energy before the nuclei stabilize at the distance r_0. Finally, a photon of fluorescence may be emitted, shown by the red vertical arrow pointing down. Note that this arrow does not reach the ground state curve at its lowest point; thus an excess of energy needs be dissipated by back-and-forth, thermal vibrations around r_0 or colli-sions with other molecules.

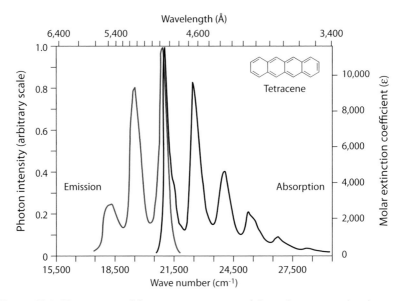

Figure 11.3. Absorption and fluorescence spectra of the polyatomic molecule tetracene in cyclohexane. The absorption (right axis) and fluorescence (left axis) spectra are mirror images of one another because the geometries of the molecules are essentially the same in the ground and in the first excited state, and therefore the energy spacing of the vibrational states are similar. But the fluorescence spectrum (shown in red) is shifted to the red. Note that the units for the wavelength scale (*top*) are angstroms (1 A = 0.1 nm). The bottom scale, in cm^{-1}, is a linear scale, proportional to energy.

the emission spectrum of the "donor," as well as on the concentration of the acceptor. It can be represented as follows, with M the donor molecule and N the acceptor molecule:

$$M^* \rightarrow M + h\nu_M$$
$$N + h\nu_M \rightarrow N^*$$
$$N^* \rightarrow N + h\nu_N$$

However, electronic energy can also pass from one molecule to another by processes that do not involve emission/reabsorption but still require that the absorption spectrum of the acceptor overlap the emission spectrum of the donor. For such a "nonradiative" resonance process to occur, the two molecules must be close (within 50 nm), but need not actually collide. The efficiency of this type of *energy transfer,* often called Förster transfer (or FRET, for Fluorescence Resonance

Energy Transfer), depends critically on the distance between donor and acceptor and on their mutual orientation. Energy transfer can and does occur in bioluminescence, as in the case of the excitation of GFP. In that case, it is referred to as *BRET*, for Bioluminescence Resonance Energy Transfer. Jellyfish bioluminescence, emitted by their GFPs, is indeed a treasure trove of spectacular examples (Chapter 2) exploited every day by molecular biologists (see Chapter 10).*

To summarize, keeping in mind that an electronically excited molecule M* could be the product of either the absorption of a photon or of a chemical reaction, the main deactivating processes are as follows:

$$M^* \rightarrow M + heat \quad \text{deactivation}$$
$$M^* \rightarrow M + h\nu_M \quad \text{emission of fluorescence of M}$$
$$M^* + N \rightarrow M + N^* \quad \text{resonance energy transfer}$$
$$N^* \rightarrow N + h\nu_N \quad \text{emission of fluorescence of N}$$
$$M^* (+ X) \rightarrow products \quad \text{photochemical reaction}$$

In this list, the excited molecules were all assumed to be in the singlet state, i.e., with all electrons spin-paired. One more process, alluded to in Figure 11.1 (right side), must be added. In some cases, the direction of spin of the electron promoted to the excited S_1 state may be reversed on a very fast time scale. This is determined by the structure of a molecule, such as the presence of a carbonyl group ($>C=O$), as in acetone ($CH_3-CO-CH_3$). As a result, the two outermost electrons of such an excited molecule are now "unpaired," spinning in the same direction, and putting the molecule in a "triplet" state (T_1); the superscripts 1 or 3 are the shorthand symbols for singlet or triplet.

$$^1M^* \rightarrow {}^3M^* \quad \text{intersystem crossing}$$
$$^3M^* \rightarrow {}^1M + h\nu \quad \text{phosphorescence}$$

---

* Non-resonance energy transfer can also occur if donor and acceptor molecules get in close contact; this process (called collisional energy transfer) is much less selective. It can take place between a triplet excited donor and some acceptor.

This unpaired electron cannot fall back to the ground state without reversing the direction of its spin. Because this is a slow process, triplet excited states have longer lifetimes (microseconds, or even longer in the gas phase at low pressure) than singlet excited states ($\sim 10^{-8}$-10^{-9} s), which makes triplet states prone to deactivation via collisions.

Molecular oxygen is a particularly efficient and important quencher of triplet excited molecules.

$$^3M^* + {}^3O_2 \; \rightarrow \; {}^1M + {}^1O_2 + h\nu \quad \text{(infrared)}$$
$$\rightarrow$$

As mentioned earlier, the ground state of oxygen is a triplet state, $^3O_2$. In quenching a triplet-state molecule, oxygen becomes excited to its singlet state, $^1O_2$, a low-lying excited state with very selective reactivity and important roles in chemistry and biology; however, there is no direct indication at this time that "singlet oxygen" plays a role in bioluminescence.

If triplet excited molecules, $^3M^*$, do eventually emit, this emission is, by definition, a phosphorescence (Figure 11.1). The needles of our clocks are often painted with phosphorescent compounds, such as salts of europium, which store the light energy they were exposed to during the day and re-emit it slowly at night. Singlet–singlet transitions are said to be "allowed," and triplet–singlet transactions "forbidden." (In photochemists' language, "forbidden" translates to slow or very slow). Since our focus is on bioluminescence emissions that are intense enough to be readily seen at night, all are likely to be $S_1 \rightarrow S_0$ transitions akin to fluorescence, rather than to phosphorescence.

Light from Chemical and Biochemical Reactions

In bioluminescence, light is *always* the product of a chemical reaction, and all well-understood bioluminescences, such as those discussed in Part I, are enzyme-catalyzed reactions involving molecular oxygen, represented below, where luciferin and luciferase are generic:

$$O_2, \text{luciferase}$$

Luciferin	$\rightarrow$	oxidized luciferin*
		(+ other products)
oxidized luciferin*	$\rightarrow$	oxidized luciferin + hν

The efficiency of a bioluminescence, *its quantum yield*, is defined as the number of photons emitted per molecule of luciferin oxidized; it is always less than 1.

Most bioluminescence emissions are in the range of blue to yellow (450–580 nm). Photons of this color carry large amounts of energy, ~52 to 65 kcal/mole (i.e., 7 to 9 times as much energy as released by the hydrolysis of ATP). Importantly, this energy must come from a single chemical step, as opposed to a series of successive chemical events whose overall energetic balance may be positive by the same amount. The question, then, arises: what sort of chemical processes can one expect to deliver such large energies in one step? The short answer is, very few. But the requirement for oxygen in bioluminescence gives us a clue. Oxygen generates peroxides, molecules carrying an oxygen-oxygen bond (as in hydrogen peroxide HOOH). Peroxides are indeed key intermediates in the chemistry of bioluminescence and of chemiluminescence in solution. The ubiquitous "light stick," the most famous byproduct of fundamental research on chemiluminescence, is a case in point (see below).

Why peroxides? Primarily, because the O-O bond of peroxides is a weak bond compared to bonds between carbon and hydrogen, or carbon and oxygen, or between two carbon atoms, or, especially, compared to a double-bonded carbonyl C=O. It takes little expense of energy to break an O-O bond; yet once it is broken, the two oxygens, now separated, are ready to form two new strong carbon-oxygen bonds.

Consider, for example, the small cyclic, peroxides, called *dioxetanes* (Figure 11.4). The square peroxide ring is stressed, which means

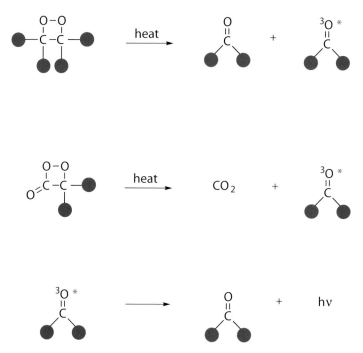

Figure 11.4. Generation of electronically excited products from four-member ring di-oxetanes. The blue circles represent substituent groups such as methyl or other ali-phatic groups. When heated, the weak O-O bond breaks first, releasing the ring strain. The C-C bond then breaks, generating two carbonyl products and enough energy for one of these products to be in an excited state, here the triplet state. Simple *dioxeta-nones* undergo a similar breakdown to generate a carbonyl product, some of it in its triplet state, and carbon dioxide. In both cases, little light is emitted because the ex-cited products are in their metastable triplet states, which are poor emitters.

that a significant amount of energy is immediately released when the weak O-O bond breaks. This leads, in short order, to cleavage of the C-C bond and to the formation of two new carbonyl C=O bonds, with the release of enough energy to generate one of the new carbonyl products in an excited state. This is indeed the scenario observed in the case of the many dioxetanes that have been synthesized in the laboratory and studied in solution. All do come apart when heated, all generate excited carbonyl products, and all emit light—though, disap-pointingly, it is only very dim light. This is because the excited prod-ucts are, by and large, in their triplet rather than their singlet excited state, and hence they are poor emitters. Note, however, that there is evidence that molecules in enzymatically generated triplet states may play important biological roles. (Since these are "dark" processes, they are not discussed here).

Similarly, dioxetanones break down to CO_2 and a product molecule carrying a carbonyl group; dioxetanones, transiently generated in these reactions, are the energy-rich molecular species precursors of the luminescence emission of cypridinids, coelenterates, and the fireflies (Chapters 1, 2, and 3).

It turns out that the yield of singlet excited products, and thus the light yield from the breakdown of a dioxetane, depends critically on the specific substituents on the dioxetane ring. Remarkably, if one of these substituents is an aromatic group (such as a benzene derivative) that can be ionized and thus made to carry a negative charge, the course of the reaction changes drastically. An intramolecular charge transfer of this negative charge to the peroxidic O-O bond, as schematized in Figure 11.5, results in the immediate decomposition of the dioxetane with a brief and intense flash of light.

One of the carbon atoms of the dioxetane ring is substituted by a bulky aliphatic group (an adamantyl group, shown in Figure 11.5), whose function is to stabilize the dioxetane by slowing down rotation around the C-C bond. The other carbon atom of the dioxetane ring carries a phenolate group, OX (where X stands for a "protective" silyl group that can be removed at will by the addition of fluoride, F^-). Heated in a nonpolar solvent, such as xylene, dioxetane D behaves likes the simple dioxetanes discussed above. It generates

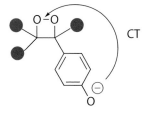

Figure 11.5. Decomposition of generic dioxetanes. Intramolecular transfer (CT) of the negative charge carried by the phenolic substituent to the O-O of a dioxetane (with neutral substituents symbolized by blue circles on the other carbon of the dioxetane ring) brings about the dioxetane decomposition. As shown below in Figure 11.6 and Figure 11.7, the course and outcome of the dioxetane decomposition depends on the solvent and the phenolic substituents.

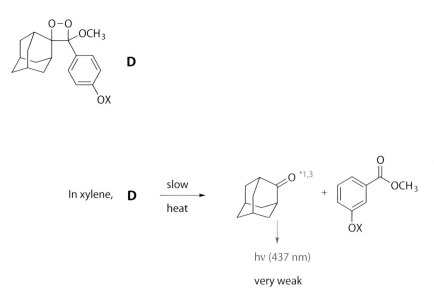

Figure 11.6. Decomposition of dioxetane D in the nonpolar solvent xylene. Some ada-
mantone is generated in its singlet excited state, resulting from a very weak chemiluminescence at 437 nm spectrally identical to the fluorescence of this product.

some adamantanone in its triplet state and a low yield of excited singlet states, therefore only a very weak emission (adamantanone fluorescence at 437 nm; Figure 11.6).

But in a polar solvent, such as acetonitrile (CH_3CN), the addition
of fluoride removes the protecting silyl group, leaving a negatively
charged phenoxide group (Figure 11.7). This brings about the immediate decomposition of D with an intense flash of light (at 466 nm,
benzoate fluorescence). Fifty percent of the benzoate is generated in
its singlet excited state.

What accounts for the drastic change in stability and light output
caused by the removal of the protecting silyl group is not yet firmly
established. It certainly starts with a transfer of negative charge
from the phenoxide -O⁻ group to the O-O bond, causing it to
break and resulting directly or indirectly (this is still debated) in
the generation of singlet excited product. Since D could have been
equally well protected by an enzymatically removable group (such
as a phosphate, removable by alkaline phosphatase), applications are
numerous.

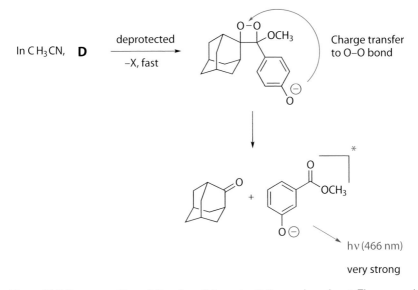

In CH₃CN, **D** → deprotected, −X, fast

Charge transfer to O–O bond

O–O ... OCH₃

hν (466 nm)

very strong

Figure 11.7. Decomposition of dioxetane D in acetonitrile, a polar solvent. The removal of the protecting group X (such as a silyl group removable by addition of fluoride) creates a negative charge on this carbon atom and results in the immediate transfer of this charge to the O-O bond of the dioxetane ring. This brings about its immediate decomposition with intense emission of benzoate fluorescence at 466 nm.

The decomposition of transient dioxetanones in the bioluminescence pathways of cypridinids (Chapter 1), coelenterates (Chapter 2), and the fireflies (Chapter 3) is likely to also involve motion of electric charges favoring the generation of singlet excited products, as in the case of the model dioxetane *D* (see below).

In the meantime, chemists have succeeded in designing model systems that rival bioluminescence in the efficiency with which they generate excited products nonenzymatically. The all-familiar "light stick" (Figure 11.8) is both a compliment to chemists and a lesson in modesty: the precise mechanism leading to the luminescence has not yet been fully sorted out, although the overall chemistry is certainly parallel to that of dioxetane *D* presented above. We now understand what is required for emulating the high efficiency of systems such as that of the fireflies: first, a reaction exothermic enough to potentially generate visible photons in a single reaction step, second, a reaction that, when triggered, is fast enough to result in the emission of a flash of light, as opposed to a long-lasting but faint

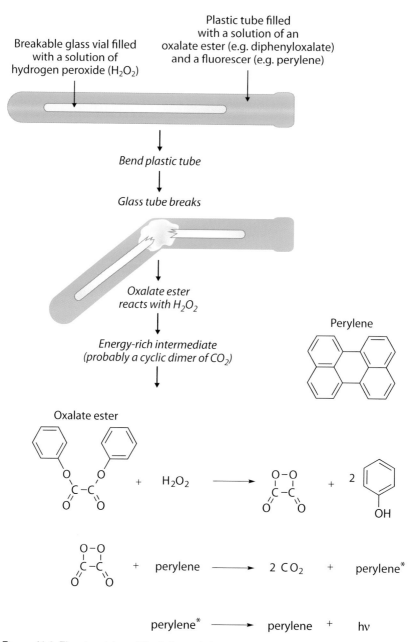

Figure 11.8. The chemistry of the light stick. It involves hydrogen peroxide, an ester of oxalic acid, such as diphenyl ester, and a good fluorescer and electron donor (sometimes called an activator), represented here by perylene. The critical energy-rich intermediate, the "dimer" of CO_2, has not been isolated.

glow, and finally, a reaction that generates an electronically excited product with a high fluorescence quantum yield.

The Chemistry of Bioluminescence Reactions

How does our present knowledge of the mechanisms of some bioluminescence reactions fit into this model? In truth, remarkably well in the cases of cypridinids, coelentereates, and the fireflies. The case of the cypridinid *Vargula* makes this point. Bioluminescence in this taxon was established to involve a transient peroxide intermediate of cyclic structure, too unstable to isolate; its breakdown is the source of the energy required to generate the product oxyluciferin in its excited state, hence the bioluminescence (see Box 1.2). That page of research on the mechanism of bioluminescence owes much to chemists, who established the role of a transient dioxetanone in the pathway to light emission.

Since the coelenterate systems of *Renilla* and *Aequorea* luciferins share with cypridinids the critical imidazolopyrazinone structure, one can safely assume that their bioluminescence reactions also proceed through similar dioxetanone intermediates. And so does the bioluminescence reaction of *fireflies*, where the role of a dioxetanone intermediate as the source of the energy released by the emission photons has also been established (see Box 3.1).

In the bacterial and dinoflagellate examples of bioluminescence systems, also discussed in Part I, there is no evidence thus far pointing to the role of dioxetanones. In the bacterial bioluminescence reaction, below, an intermediate flavin peroxide is formed, and the emitter is a hydroxyl derivative of the isoalloxazine part of that molecule (see Box 5.1):

$$E\text{-}FMNH_2 + O_2 + RCHO \rightarrow E\text{-}FMN + RCOOH + H_2O + h\nu$$

Regarding *dinoflagellate* bioluminescence, although the structure of the luciferin and that of the final reaction product have been established (see Chapter 4), no reaction pathway leading to excited state formation has thus far been proposed.

We hope that our readers take home the message that, little by little, biologists and chemists working together are solving the mystery of how the firefly and lowly bacteria can manage to emit an intense light without themselves getting hot in the process. The field of bioluminescence is now a collection of fruitful collaborations between scientists of very different backgrounds.

GLOSSARY

amino acids The building blocks of proteins. Their properties vary with the nature of the group represented by R in the general formula $H_2N\text{-}CHR\text{-}CO_2H$. They link to each other by the so-called peptide bond to form proteins. The nature of the group R determines the specific property of an amino acid. There are twenty common amino acids.

anoxygenic photosynthesis The process whereby light energy is captured and stored as ATP without the production of oxygen. This was one of the ways by which organisms obtained energy prior to the appearance of oxygen on earth. (see oxygenic photosynthesis)

aposematism A signal, such as a bright color, smell, or bioluminescence emission, exhibited by an organism to warn potential predators that the displaying organism would be an unfavorable prey; the avoidance of an attack is considered to be advantageous to both.

ATP Adenosine triphosphate (ATP) is an "energy-rich" molecule because it contains two high-energy phosphoanhydride bonds linking three phosphate groups. The hydrolysis of ATP to adenosine diphosphate (ADP) releases 7.3 kcal of energy, as does that of ADP to adenosine monophosphate. ATP and ADP are continuously recycled in living animals and provide the energy for the synthesis of macromolecules.

autotrophic organisms Organisms that obtain their energy for growth and production of complex organic molecules from either light (*photoautotrophs*) or inorganic chemical reactions (*chemoautotrophs*).

circadian rhythm The property of some biological processes to oscillate in time with a period of about 24 hours when kept in constant conditions of light, temperature, and other factors.

chromophores Molecules that absorb visible light and so are colored. They may be attached to a larger molecule such as a protein. If a chromophore re-emits the light absorbed as fluorescence, it may be called a fluorophore.

counterillumination A method of camouflage in which light emission by an organism is used to match or disrupt the image of the organism's silhouette; the ventral light emission by the *Euprymna* squid matches the intensity of moonlight, making it less visible to predators from below.

dioxetanes and dioxetanones Cyclic peroxides (made of two adjoining oxygen atoms bound to two carbon atoms). They are hypothesized to be the precursors of the light emitters in several bioluminescences, such as that of cypridinids, coelenterates and beetles.

energy transfer A process that transfers the electronic excitation of one molecule (the donor) to another (the acceptor) when these two molecules are close, but not in actual contact. See *GFPs.*

enzymes Proteins that function as biological catalysts, speeding up a reaction without changing its products; enzymes' structures are different and specific for each reaction.

fluorescence The light emitted by a molecule that had been put in an "electronic excited state" by absorbing light. The re-emission is delayed, typically by about one or several nanoseconds (billionths of a second). Fluorescence is shifted to the red of the exciting light, some of the energy loss being accounted for by vibrations and rotation of the molecule. (See also phosphorescence).

fluorescence proteins Proteins with a chromophore that associate with luciferase/luciferin systems. Their presence shifts the emission spectra of these systems; see GFPs and YFP in Chapters 2 and 5

GFPs See *green fluorescence proteins.*

green fluorescence proteins (GFPs) GFPs are highly fluorescent molecules present in coelenterates and responsible for their green emission via energy transfer from the excited reaction product (coelenteramide).

luciferin A generic term for the substrate in a luciferase reaction, in which a reaction intermediate or product may serve as the emitter. If there are two or more substrates, the one implicated as the emitter or giving rise to it is called the luciferin. Each must be identified by the species or group to which it belongs–e.g., firefly luciferin or *Latia* luciferin–or by its chemical name. Structures of several different luciferins have been reported.

luciferase A generic term for the enzyme in a bioluminescence reaction, usually identified by the species or group to which it belongs. Luciferases from different organisms typically have no structural similarities and their genes are unrelated–e.g., bacterial luciferase or *Renilla* luciferase.

lux operon A cluster of coregulated genes involved in bacterial bioluminescence.

oxygenic photosynthesis The process by which light splits water into electrons, protons, and free oxygen. It evolved sometime prior to 2.5 million years ago (see anoxygenic photosynthesis).

peroxides Molecules in which two oxygen atoms are linked together to form a peroxidic bond R-O-O-R. The simplest is *hydrogen peroxide*, H-O-O-H. The O-O bond is a weak bond, compared to a -C-O- or -C-H bond, and therefore easy to break. This can release a large amount of energy if stronger bonds get formed as the result of breaking the O-O bond. Peroxides are the source of the energy released as photons of blue-green light in the bioluminescence reactions of bacteria, cnidarians, coelenterates, and beetles.

phosphorescence Similar to fluorescence, it also results from prior absorption of light, but the re-emission is delayed longer, sometimes persisting up to days, because the molecule is put in a metastable state. In biological conditions, durations longer than tens or hundreds of nanoseconds, perhaps microseconds, are unlikely.

photocyte A cell that emits light.

photons Particles (or quanta) of energy carried by light of a given color (wavelength), according to Einstein's theory. Each photon carries an amount of energy equal to $h\nu$, where h is the Planck constant and ν the frequency of the light (inversely proportional to its wavelength). A photon of green light, at 530 nm, carries about 53 kcal/mole.

photophore A light-emitting organ, comprising many photocytes.

proteins Chains of many—usually hundreds—of amino acids (sometimes called residues) linked together in a specific order by the peptide bond between the CO_2H group of one amino acid and the NH_2 group of another (as here between two amino acids, NH_2-CHR-CO-NH-CHR-CO_2H).

quorum sensing A mechanism of chemical communication between bacteria mediated by an "autoinducer," a substance produced by the bacteria themselves. At a critical concentration of the autoinducer in the medium, gene expression is induced; thus, the autoinducer serves as proxy for the bacterial concentration.

regulon A collection of genes or operons under regulation by the same regulatory molecule or mechanism; generally used for prokaryotic systems—for example, quorum sensing in bacteria.

reporters Genes with a readily detectable product, such as bioluminescence. Attached to the regulatory sequence of a gene of interest, the time and location of the expression of that gene can be determined.

symbiosis Any relationship between two individuals of different species that is beneficial to both.

singlet and triplet states In most molecules in their lowest energy state (the ground state), all electrons are paired, i.e., one spinning in one direction, the other in the opposite direction. This is also the case for the lowest excited electronic state. When such molecules fall from their excited singlet state back to the ground state, there is no change in the direction of electron spin and the emission is a fluorescence. If a change of spin direction is necessary, the triplet state of the molecule is formed and the emission is a phosphorescence, a slower process.

Greek Letters

μ (mu) The symbol for micro, or one-millionth.

λ (lambda) Refers to wavelength in nanometers (nm) of electromagnetic radiation (light).

ν (nu) The symbol for the frequency of light in wave numbers (reciprocal centimeters, cm^{-1}).

Units and Prefixes

The units of time in this book are hours, minutes, and seconds. In the case of seconds, the prefixes m, μ, and n, which stand for milli, micro, and nano, respectively, are fractions of seconds: a ms is one-thousandth of a second, a μs is one-millionth of a second, and a ns is one-millionth of a ms, or a billionth of a second.

The units of distance are the meter, the cm, the mm, and the μ (or μm). The angstrom, Å, sometimes but rarely used here, is equal to 0.1 mμ.

FURTHER READING

Web

The Web has become a rich source of photos and videos of bioluminescent organisms. A website maintained by the American Society for Photobiology, Photobiological Sciences Online, has several good articles on bioluminescence of a number of different organisms: http://photobiology.info/#Biolum.

General References

Newton Harvey's book is the classic text; it describes most organisms known to emit bioluminescence. It truly put bioluminescence on the map of academic research. After 60 years, it remains an invaluable resource. His later publication on the history of bioluminescence before 1900 is extremely interesting.

Another important book, edited by P. J. Herring, collects chapters authored by experts in their fields on different aspects of bioluminescence, from photophysics to evolution.

Pierebone and Gruber's book is a lovely presentation on bioluminescence. It introduces readers to some of the scientists who carry on the work, as well as to organisms they selected for study, and it explains to the lay reader, in simple terms, how bioluminescence works. The book by O. Shimomura is a gold mine of very specific information on the biochemistry of bioluminescence.

Harvey, E. N. 1952. *Bioluminescence*. Academic Press, New York.
Harvey, E. N. 1957. "A History of Luminescence from the Earliest Times until 1900." *Memoirs of the American Philosophical Society* 44:1–692.
Hastings, J. W. 2012. "Bioluminescence." In *Cell Physiology Sourcebook: Essentials of Membrane Biophysics*, 4th ed., ed. N. Sperelakis, pp. 925–947. Academic Press, New York.
Herring, P. J., ed. 1978. *Bioluminescence in Action*. Academic Press, New York.

Pieribone, V., and D. F. Gruber. 2005. *Aglow in the Dark.* The Belknap Press of Harvard University Press, Cambridge.

Shimomura, O. 2006. *Bioluminescence: Chemical Principles and Methods.* World Scientific Publishing Co., Singapore.

Wilson, T., and J. W. Hastings. 1998. "Bioluminescence." *Annual Review of Cell Developmental Biology* 14:197–230.

Additional Reading by Chapter

1. A Marine Crustacean

Haneda, Y., and F. H. Johnson. 1962. "Photogenic organs of *Parapriacanthus beryciformes Franz* and other fish with the indirect type of luminescent system." *Journal of Morphology* 110:187–198.

Kato, S., Y. Oba, M. Ojika, and S. Inouye. 2004. "Identification of the biosynthetic units of *Cypridina* luciferin in *Cypridina (Vargula) hilgendorfii* by LC/ESI-TOF-MS." *Tetrahedron* 60:11427–11434.

Morin, J. G., and A. C. Cohen. 2010. "It's all about sex: Bioluminescent courtship displays, morphological variation, and sexual selection in two new genera of Caribbean ostracodes." *Journal of Crustacean Biology* 30:56–67.

Rivers, T. J., and J. G. Morin. 2008. "Complex sexual courtship displays by luminescent male marine ostracods." *Journal of Experimental Biology* 211:2252–2262.

Thompson, E. M, and F. I. Tsuji. 1989. "Two populations of the marine fish *Porichthys notatus,* one lacking in luciferin essential for bioluminescence." *Marine Biology* 102:161–165.

2. Jellyfish and Green Fluorescent Protein

Chalfie, M., and S. R. Kain, eds. 2006. *Green Fluorescent Protein: Properties, Applications, and Protocols.* Wiley-Interscience, Hoboken.

Haddock, S. H. D., T. J. Rivers, and B. H. Robison. 2001. "Can coelenterates make coelenterazine? Dietary requirement for luciferin in cnidarian bioluminescence." *Proceedings of the National Academy of Sciences (USA)* 98:11148–11152.

Inoue, S., S. Sugiura, H. Kakoi, K. Hasizume, T. Goto, and H. Iio. 1975. "Squid Bioluminescence II. Isolation from *Watasenia scintillans* and synthesis of 2-(p-hydroxybenzyl)-6-(p-hydroxyphenyl)-3,7-dihydroimidazo(1,2-a)pyrazine-3-one." *Chemistry Letters* 4(2):141–144.

Rees, J-F., B. De Wergifosse, O. Noiset, M. Dubuisson, B. Janssens, and E. M. Thompson. 1998. "The origins of marine bioluminescence: Turning oxygen defence mechanisms into deep-sea communication tools." *Journal of Experimental Biology* 201:1211–1221.

Shimomura, O. 2005. "The discovery of aequorin and green fluorescent protein." *Journal of Microscopy* 217:3–15.

3. Fireflies and Other Beetles

Ando, Y., K. Niwa, N. Yamada, T. Enomoto, T. Irrie, H. Kubota, Y. Ohmiya, and H. Akiyama. 2008. "Firefly bioluminescence quantum yield and color change by pH-sensitive green emission." *Nature Photonics* 2:44–47.

Conti, E., N. P. Franks, and P. Brick. 1996. "Crystal structure of firefly luciferase throws light on a superfamily of adenylate-forming enzymes." *Structure* 4:287–298.

Dubois, R. 1886. "Les Elatérides Lumineux." *Bulletin Société Zoologie de France* vol. xi, pp. 1–275

Fraga, H. 2008. "Firefly luminescence: A historical perspective and recent developments." *Photochemical and Photobiological Sciences* 7:146–158.

Nakatsu, T., S. Ichiyama, J. Hiratake, A. Saldanha, N. Kobashi, K. Sakata, and H. Kato. 2006. "Structural basis for the spectral difference in luciferase bioluminescence." *Nature* 440:372–376.

Viviani, V. R. 2002. "The origin, diversity, and structure function relationships of insect luciferases." *Cellular and Molecular Life Sciences* 59:1833–1850.

Wood, K. V. 1995. "The chemical mechanism and evolutionary development of beetle bioluminescence." *Photochemistry and Photobiology* 62:662–673.

4. Dinoflagellates and Krill

Heimann, K., P. L. Klerks, and K. H. Hasenstein. 2009. "Involvement of actin and microtubules in regulation of bioluminescence and translocation of chloroplasts in the dinoflagellate *Pyrocystis lunula*." *Botanica Marina* 52:170–177.

Latz, M. I., J. Allen, S. Sarkar, and J. Rohr. 2009. "Effect of fully characterized unsteady flow on population growth of the dinoflagellate *Lingulodinium polyedrum*." *Limnology and Oceanography* 54:1243–1256.

Latz, M. I., M. Bovard, V. VanDelinder, E. Segre, J. Rohr, and A. Groisman. 2008. "Bioluminescent response of individual dinoflagellate cells to hydrodynamic stress measured with millisecond resolution in a microfluidic device." *Journal of Experimental Biology* 211:2865–2875.

Mensinger, A. F., and J. F. Case. 1992. "Dinoflagellate luminescence increases susceptibility of zooplankton to teleost predation." *Marine Biology* 112:207–210.

Smith S. M. E., D. Morgan, B. Musset, V. V. Cherny, A. R. Place, J. W. Hastings, and T. E. DeCoursey. 2011. "Voltage-gated proton channel in a dinoflagellate." *Proceedings of the National Academy of Sciences (USA)* 108:18162–18167.

Taylor, F. J. R., ed. 1987. *The Biology of Dinoflagellates*. Blackwell Scientific Publications, Oxford.

5. Bacteria

Hastings, J. W. 1971. "Light to hide by: Ventral luminescence to camouflage the silhouette." *Science* 173:1016–1017.

Meighen, E.A. 1991. "Molecular biology of bacterial bioluminescence." *Microbiological Reviews* 55:123–142.

Miller, S. D., S. H. D. Haddock, C. D. Elvidge, and T. F. Lee. 2005. "Detection of a bioluminescent milky sea from space." *Proceedings of the National Academy of Sciences (USA)* 102:14181–14184.

Ng, W. L., and B. L. Bassler. 2009. "Bacterial quorum-sensing network architectures." *Annual Review of Genetics* 43:197–222.

Zarubin, M., S. Belkin, M. Ionescu, and A. Genin. 2012. "Bacterial bioluminescence as a lure for marine zooplankton and fish." *Proceedings of the National Academy of Sciences (USA)* 109:853–857.

6. Short Accounts of Other Luminous Organisms

Claes, J. M., D. L. Aksnes, and J. Mallefet. 2010. "Phantom hunter of the fjords: Camouflage by counterillumination in a shark *(Etmopterus spinax)*." *Journal of Experimental Marine Biology and Ecology* 388:28–32.

Deheyn, D. D., and M. I. Latz. 2009. "Internal and secreted bioluminescence of the marine polychaete *Odontosyllis phosphorea* (Syllidae)." *Invertebrate Biology* 128:31–45.

Desjardin, D. E., A. G. Oliveira, and C. V. Stevani. 2008. "Fungi bioluminescence revisited." *Photochemical & Photobiological Sciences* 7:170–182.

Goodrich-Blair, H., and D. J. Clarke. 2007. "Mutualism and pathogenesis in *Xenorhabdus* and *Photorhabdus*: Two roads to the same destination." *Molecular Microbiology* 64:260–268.

Henry, J. P., and C. Monny. 1977. "Protein-protein interactions in the *Pholas dactylus* system of bioluminescence." *Biochemistry* 16:2517–2525.

Kuse, M., E. Tanaka, and T. Nishikawa. 2008. "Pholasin luminescence is enhanced by addition of dehydrocoelenterazine." *Bioorganic and Medicinal Chemistry Letters* 18:5657–5659.

Lehtiniemi, M., A. Lehmann, J. Javidpour, and K. Myrberg. 2012. "Spreading and physico-biological reproduction limitations of the invasive American comb jelly *Mnemiopsis leidyi* in the Baltic Sea." *Biological Invasions* 14:341–354.

Marek, P., D. Papaj, J. Yeager, S. Molina, and W. Moore. 2011. "Bioluminescent aposematism in millipedes." *Current Biology* 21:R680–R681.

Merritt, D. J., and A. K. Clarke. 2011. "Synchronized circadian bioluminescence in cave-dwelling *Arachnocampa tasmaniensis* (glowworms)." *Journal of Biological Rhythms* 26:34–43.

Meyer-Rochow, V. B. 2007. "Glowworms: A review of *Arachnocampa* spp. and kin." *Luminescence* 22:251–265.

Meyer-Rochow, V. B., and M. V. Bobkova. 2001. "Anatomical and ultrastructural comparison of the eyes of two species of aquatic, pulmonate gastropods: The bioluminescent *Latia neritoides* and the non-luminescent *Ancylus fluviatilis*." *New Zealand Journal of Marine and Freshwater Research* 35:739–750.

Morin, J. G., A. Harrington, N. Krieger, K. H. Nealson, T. O. Baldwin, and J. W. Hastings. 1975. "Light for all reasons: Versatility in the behavioral repertoire of the flashlight fish." *Science* 190:74–76.

Ohmiya, Y., S. Kojima, M. Nakamura, and H. Niwa. 2005. "Bioluminescence in the limpet-like snail, *Latia neritoides*." *Bulletin of the Chemical Society of Japan* 78:1197–1205.

Oliveira, A. G., and C. V. Stevani. 2009. "The enzymatic nature of fungal bioluminescence." *Photochemical & Photobiological Sciences* 8:1416–1421.

Rudie, N. G., M. G. Mulkerrin, and J. E. Wampler. 1981. "Earthworm bioluminescence: Characterization of high specific activity *Diplocardia longa* luciferase and the reaction it catalyzes." *Biochemistry* 20:344–350.

Viviani, V. R., J. W. Hastings, and T. Wilson. 2002. "Two bioluminescent diptera: The North American *Orfelia fultoni* and the Australian *Arachnocampa flava*. Similar niche, different bioluminescence systems." *Photochemistry and Photobiology* 75:22–27.

7. Bioluminescence in the Oceans

Bowmaker, J. K., V. I. Govardovskii, S. A., Shukolyukov, L. V. Zueva, D. M. Hunt, V. G. Sideleva, and O. G. Smirnova. 1994. "Visual pigments and the photic environment—the cottoid fish of Lake Baikal." *Vision Research* 34:591–605.

Douglas R. H., J. C. Partridge, K. S. Dulai, D. M. Hunt, C. W. Mullineaux, and P. H. Hynninen. 1999. "Enhanced retinal longwave sensitivity using a chlorophyll-derived photosensitiser in *Malacosteus niger*, a deep-sea dragon fish with far red bioluminescence." *Vision Research* 39:2817–2832.

Douglas, R. H., J. C. Partridge, and N. J. Marshall. 1998. "The eyes of deep-sea fish I: Lens pigmentation, tapeta, and visual pigments." *Progress in Retinal Eye Research* 17:597–636.

Herring, P. J., and C. Cope. 2005. "Red bioluminescence in fishes: On the suborbital photophores of *Malacosteus*, *Pachystomias* and *Aristostomias*." *Marine Biology* 148:383–394.

Pietsch, T. W. 2009. *Oceanic Anglerfishes: Extraordinary Diversity in the Deep-sea*. University of California Press, Berkeley and Los Angeles.

Warrant, E. J., and N. A. Locket. 2004. "Vision in the deep sea." *Biological Reviews* 79:671–712.

Widder, E. A. 2010. "Bioluminescence in the ocean: Origins of biological, chemical, and ecological diversity." *Science* 328:704–708.

8. The Many Functions of Bioluminescence

Harper, R. D., and J. F. Case. 1999. "Disruptive counterillumination and its anti-predatory value in the plainfin midshipman *Porichthys notatus*." *Marine Biology* 134:529–540.

Hastings, J. W. 1971. "Light to hide by: Ventral luminescence to camouflage the silhouette." *Science* 173:1016–1017.

Hastings, J. W. 1983. "Biological diversity, chemical mechanisms, and evolutionary origins of bioluminescent systems." *Journal of Molecular Evolution* 19:309–321.

Morin, J. G. 1983. "Coastal bioluminescence: Patterns and functions." *Bulletin of Marine Science* 33:787–817.

Willis, R. E., C. R. White, and D. J. Merritt. 2011. "Using light as a lure is an efficient predatory strategy in *Arachnocampa flava*, an Australian glowworm." *Journal of Comparative Physiology B* 181:477–486.

9. The Origins and Evolution of Bioluminescence

Dahl, T. W., E. U. Hammarlund, A. D. Anbar, D. P. G. Bond, B. C. Gill, G. W. Gordon, A. H. Knoll, A. T. Nielsen, N. H. Schovsbo, and D. E Canfield. 2010. "Devonian rise in atmospheric oxygen correlated to the radiations of terrestrial plants and large predatory fish." *Proceedings of the National Academy of Sciences (USA)* 107:17911–17915.

Holland, H. D. 2006. "The oxygenation of the atmosphere and oceans." *Philosophical Transactions of the Royal Society B* 361:903–915.

Holland H. D. 2009. "Why the atmosphere became oxygenated: A proposal." *Geochimica et Cosmochimica Acta* 73:5241–5255.

McElroy, W. D., and H. H. Seliger. 1962. "Origin and evolution of bioluminescence." In *Horizons in Biochemistry*, ed. M. Kasha and B. Pullman, pp. 91–101. Academic Press, New York.

10. Applications

Alves, E, L. Costa, A. Cunha, M. A. F. Faustino, M. Neves, and A. Almeida. 2011. "Bioluminescence and its application in the monitoring of antimicrobial photodynamic therapy." *Applied Microbiology and Biotechnology* 92:1115–1128.

Hadjantonakis, A. K., S. Macmaster, and A. Nagy. 2002. "Embryonic stem cells and mice expressing different GFP variants for multiple non-invasive reporter usage within a single animal." *BMC Biotechnology* 2:11.

Kondo, T., C. A. Strayer, R. D. Kulkarni, W. Taylor, M. Ishiura, S. Golden and C. H. Johnson. 1993. "Circadian rhythms in prokaryokes: Luciferase as a reporter of circadian gene expression in cyanobacteria." *Proceedings of the National Academy of Sciences (USA)* 90:5672–5676.

Lichtman, J. W., J. Livet, and J. R. Sanes. 2008. "Progress—A technicolour approach to the connectome." *Nature Reviews Neuroscience* 9:417–422.

Livet J., T. A. Weissman, H. N. Kang, R. W. Draft, J. Lu, R. A. Bennis, J. R. Sanes, and J. W. Lichtman. 2007. "Transgenic strategies for combinatorial expression of fluorescent proteins in the nervous system." *Nature* 450:56–62.

Niell, C. M., M. P. Meyer, and S. J. Smith. 2004. "In vivo imaging of synapse formation on a growing dendritic arbor." *Nature Neuroscience* 7:254–260.

Roda, A., ed. 2010. *Chemiluminescence and Bioluminescence: Past, Present and Future.* Royal Society of Chemistry, London.

Watanabe, Y., and Y. Tanaka. 2011. "Bioluminescence-based imaging technique for pressure measurement in water." *Experiments in Fluids* 51, 225–236.

Wu, C., K. Mino, H. Akimoto, M. Kawabata, K. Nakamura, M. Ozaki, and Y. Ohmiya. 2009. "In vivo far-red luminescence imaging of a biomarker based on BRET from *Cypridina* bioluminescence to an organic dye." *Proceedings of the National Academy of Sciences (USA)* 106:15599–15603.

11. How Does Life Make Light?

Baader, W. J., C. V. Stevani, and E. L. Bastos. 2006. "Chemiluminescence of organic peroxides." In *The Chemistry of Peroxides*, ed. Z. Rappoport, pp. 1211–1278. John Wiley, Marblehead.

Berlman, I. B. 1965. *Handbook of Fluorescence Spectra of Aromatic Molecules.* Academic Press, New York.

Engel, T., and P. Reid. 2006. *Physical Chemistry.* Pearson Benjamin-Cummings, San Francisco.

Halpern, A. M. 2005. "Luminescence (fluorescence and phosphorescence)." In *The Encyclopedia of Physics*, ed. R. G. Lerner and G. L. Trigg, pp. 1349–1351. Wiley-VCH, New York.

Turro, N. J., V. Ramamurthy, and J. C. Scaiano. 2009. *Principles of Molecular Photochemistry—An Introduction.* University Science Books, Sausalito.

Wardle, B. 2010. *Principles and Applications of Photochemistry.* John Wiley and Sons, Marblehead.

Wilson, T. 1995. "Comments on the mechanisms of chemi- and bioluminescence." *Photochemistry and Photobiology* 62:601–606.

In addition, Figures 11.6 and 11.7 are based on years of pioneering work by A.P. Schaap, W. Adam, I. Bromstein, R. F. Vassil'ev, M. Matsumoto, and A.V. Trofimov.

ILLUSTRATION CREDITS

Every effort has been made to obtain permission for copyrighted images used in this book. Please contact the authors or publisher as needed so that any corrections may be included in a future printing of this work.

Page i: Courtesy of J. Michael Sauder, Carlsbad, Calif.

Introduction

Figure I.1: Courtesy of J. Michael Sauder, Carlsbad, Calif.

Chapter 1

Figure 1.1A: Courtesy of O. Shimomura, Marine Biological Laboratory.
Figure 1.1B: Courtesy of Martin Dohrn.
Figure 1.2: Courtesy of Martin Dohrn.
Figure 1.3: Courtesy of James Morin, Cornell University, Ithaca, N.Y.
Figure 1.5: *A.* Courtesy of Robert Lee. *B.* Courtesy of Milton Cormier, University of Georgia.
Figure 1.6: From Y. Haneda and F. H. Johnson, 1962, "Photogenic organs of *Parapriacanthus beryciformes Franz* and other fish with the indirect type of luminescent system," *Journal of Morphology* 110:187–198.

Chapter 2

Figure 2.1: Courtesy of Milton Cormier, University of Georgia.
Figure 2.4: Courtesy of O. Shimomura, Marine Biological Laboratory.
Figure 2.5: From the RCSB Protein Data Bank, "Green Fluorescent Protein from *Aequorea victoria*," http://www.rcsb.org/pdb/explore/explore.do?structureId=1EMA.
Figure 2.6: Courtesy of Harvard Museum of Comparative Zoology; Andrew Williston, photographer.

Chapter 3

Figure 3.1: Courtesy of Arwin Provonsha, Purdue University.
Figure 3.3: Modified from Y. Ando, K. Niwa, N. Yamada, T. Enomoto, T. Irrie, H. Kubota, Y. Ohmiya, and H. Akiyama, 2008, "Firefly bioluminescence quantum yield and color change by pH-sensitive green emission," *Nature Photonics* 2:44–47.

Figure 3.5: By Sze-Yi Lau from open source coordinates. N. P. Franks, A. Jenkins, E. Conti, W. R. Lieb, and P. Brick, *Biophysical Journal* (1998): 75, 2205–2211; and E. Conti et al., *Structure* (1996) 4, 287–298.

Figure 3.6: Courtesy of Vadim Viviani, Federal University of São Carlos, Sorocaba, SP, Brazil.

Figure 3.7: Courtesy of H. Ghiradella, SUNY Albany, N.Y.

Figure 3.8: Courtesy of James Jordan, photographer.

Figure 3.9: Courtesy of Ivan Poulson, photographer.

Figure 3.10: Courtesy of Etelvino Bechara, University of Sao Paulo; Sergio Vanin, photographer.

Figure 3.11: Courtesy of Etelvino Bechara, University of Sao Paulo; Sergio Vanin, photographer.

Chapter 4

Figure 4.1: Courtesy of Lawrence Fritz, Bloomsburg State University.

Figure 4.2: Courtesy of Carl H. Johnson, Vanderbilt University.

Figure 4.3: Courtesy of M.-T. Nicolas, CNRS, Montpellier, France.

Figure 4.4: Courtesy of Lawrence Fritz, Bloomsburg State University.

Figure 4.6: Courtesy of John Dolan, CNRS, Villefranche-sur-Mer.

Figure 4.7: Modified from H. Nakamura, Y. Kishi, O. Shimomura, D. Morse, and J. W. Hastings, 1989, "Structures of dinoflagellate luciferin and its enzymatic and non-enzymatic air-oxidation products," *Journal of the American Chemical Society* 111:7607–7611.

Figure 4.8: Modified from L. Li, R. Hong, and J. W. Hastings, 1997, "Three functional luciferase domains in a single polypeptide chain," *Proceedings of the National Academy of Sciences (USA)* 94:8954–8958.

Figure 4.9: From W. Schultz, L. Liu, M. Cegielski, and J. W. Hastings, 2005, "Crystal structure of a pH-regulated luciferase catalyzing the bioluminescent oxidation of an open tetrapyrrole. *Proceedings of the National Academy of Sciences (USA)* 102:1378–1383

Figure 4.10: Modified from H. Broda, D. Brugge, K. Homma, and J. W. Hastings, 1985, "Circadian communication between unicells? Effects on period by cell-conditioning of medium," *Cell Biophysics* 8:47–67.

Figure 4.11: Courtesy of Lawrence Fritz, Bloomsburg State University.

Figure 4.12: Courtesy of Jennifer Wolny, University of South Florida.

Figure 4.13: Courtesy of Edith Widder, Ocean Research and Conservation Association, Fort Pierce, Fla.

Chapter 5

Figure 5.4: Modified from K. Nealson, T. Platt, and J. W. Hastings, 1970, "The cellular control of the synthesis and activity of the bacterial luminescent system," *Journal of Bacteriology* 104:313–322.

Figure 5.6: Courtesy of Margaret McFall-Ngai, University of Wisconsin, Madison.

Figure 5.8: J. W. Hastings.

Figure 5.9: Modified from J. W. Hastings, 1971, "Light to hide by: Ventral luminescence to camouflage the silhouette," *Science* 173:1016–1017.

Figure 5.10: Courtesy of David Powell, Monterey, Calif.

Figure 5.11: Courtesy of Martin Kessel, Hebrew University.

Figure 5.12: Courtesy of Edith Widder, Ocean Research and Conservation Association, Fort Pierce, Fla.

Figure 5.13: Courtesy of Stephen D. Miller, S. H. D. Haddock, C. D. Elvidge, and T. F. Lee, 2005, "Detection of a bioluminescent milky sea from space," *Proceedings of the National Academy of Sciences (USA)* 102:14181–14184.

Chapter 6

Figure 6.1: Courtesy of David Doubilet, photographer.

Figure 6.2: Courtesy of David Doubilet, photographer.

Figure 6.3: Courtesy of J. B. Wood, Bermuda.

Figure 6.4: Courtesy of Otto Oliveira, Universidade Federal do ABC, SP, Brazil.

Figure 6.6: Courtesy of V. B. Meyer-Rochow, Jacobs University, Bremen, Germany.

Figure 6.7: Modified from Y. Ohmiya, S. Kojima, M. Nakamura, and H. Niwa, 2005, "Bioluminescence in the limpet-like snail, *Latia neritoides*," *Bulletin of the Chemical Society of Japan* 78:1197–1205.

Figure 6.8: Courtesy of M. J. Cormier, University of Georgia.

Figure 6.9: Adapted from H. Goodrich-Blair and D. J. Clarke, 2007, "Mutualism and pathogenesis in *Xenorhabdus* and *Photorhabdus*: Two roads to the same destination," *Molecular Microbiology* 64:260–268.

Figure 6.10: Courtesy of David Merritt, University of Queensland; Anthony O'Toole, photographer.

Figure 6.11: Courtesy of Waitomo Caves.

Figure 6.12: Courtesy of Dante Fenolio, Atlanta Botanical Garden, Atlanta, Ga.

Figure 6.13: From P. Marek, D. Papaj, J. Yeager, S. Molina, and W. Moore, 2011, "Bioluminescent aposematism in millipedes," *Current Biology* 21:R680–R681.

Figure 6.14: Courtesy of Cassius V. Stevani, University of Sao Paulo, Brazil.

Chapter 7

Figure 7.2: Adapted from R. H. Douglas, J. C. Partridge, and N. J. Marshall, 1998, "The eyes of deep-sea fish I: Lens pigmentation, tapeta, and visual pigments," *Progress in Retinal Eye Research* 17:597–636.

Figure 7.3: Courtesy of J. N. Marshall, Queensland Brain Institute, University of Queensland, Brisbane, Australia.

Figure 7.4: Courtesy of Museum of Comparative Zoology, Harvard University.

Figure 7.5: Courtesy of C. Kenaly, Museum of Comparative Zoology, Harvard University.

Figure 7.6: Courtesy of Adrian Flynn, University of Queensland, Australia.

Figure 7.7: *A.* Courtesy of Peter Herring. *B.* Courtesy of Edith Widder, Ocean Research and Conservation Association, Fort Pierce, Fla. From P. J. Herring and C. Cope, 2005, "Red bioluminescence in fishes: On the suborbital photophores of *Malacosteus, Pachystomias* and *Aristostomias*," *Marine Biology* 148:383–394.

Chapter 9

Figure 9.1: Modified from H. D. Holland, 2009, "Why the atmosphere became oxygenated: A proposal," *Geochimica et Cosmochimica Acta* 73:5241–5255.

Chapter 10

Figure 10.1: Courtesy of Y. Haneda.

Figure 10.2: Adapted from T. Kondo, C. A. Strayer, R. D. Kulkarni, W. Taylor, M. Ishiura, S. Golden, and C. H. Johnson, 1993, "Circadian rhythms in prokaryokes: Luciferase as a reporter of circadian gene expression in cyanobacteria," *Proceedings of the National Academy of Sciences (USA)* 90:5672–5676.

Figure 10.3: Adapted from C. Wu, C., K. Mino, H. Akimoto, M. Kawabata, K. Nakamura, M. Ozaki, and Y. Ohmiya, 2009, "In vivo far-red luminescence imaging of a biomarker based on BRET from *Cypridina* bioluminescence to an organic dye," *Proceedings of the National Academy of Sciences (USA)* 106:15599–15603.

Figure 10.4: Adapted from A.-K. Hadjantonakis, S. Macmaster, and A. Nagy, 2002, "Embryonic stem cells and mice expressing different GFP variants for multiple non-invasive reporter usage within a single animal," *BMC Biotechnol* 2:11.

Figure 10.5: Adapted from C. M. Niell, M. P. Meyer, and S. J. Smith, 2004. "In vivo imaging of synapse formation on a growing dendritic arbor," *Nature Neuroscience* 7:254–260.

Figure 10.6: Adapted from J. W. Lichtman, J. Livet, and J. R. Sanes, 2008, "Progress—A technicolour approach to the connectome," *Nature Reviews Neuroscience* 9:417–422.

Figure 10.7: Adapted from J. Livet T. A. Weissman, H. N. Kang, R. W. Draft, J. Lu, R. A. Bennis, J. R. Sanes, and J. W. Lichtman, 2007, "Transgenic strategies for combinatorial expression of fluorescent proteins in the nervous system," *Nature* 450:56–62.

Chapter 11

Figure 11.3: Adapted from I. B. Berlman, 1965, *Handbook of Fluorescence Spectra of Aromatic Molecules*, Academic Press, New York.

INDEX